Nutrition
FOR
DUMMIES®
2ND EDITION

Nutrition
FOR
DUMMIES®
2ND EDITION

**by Nigel Denby, Sue Baic,
and Carol Ann Rinzler**

WILEY

A John Wiley and Sons, Ltd, Publication

Nutrition For Dummies®, 2nd Edition

Published by
John Wiley & Sons, Ltd
The Atrium
Southern Gate
Chichester
West Sussex
PO19 8SQ
England

E-mail (for orders and customer service enquires): cs-books@wiley.co.uk

Visit our Home Page on www.wileyeurope.com

Wiley also publishes its books in a variety of electronic formats. Some content that appears in print may not be available in electronic books.

British Library Cataloguing in Publication Data: A catalogue record for this book is available from the British Library

ISBN: 978-0-470-97276-2 (paperback), ISBN: 978-0-470-97434-6 (ebk),

ISBN: 978-0-470-97304-2 (ebk), ISBN: 978-0-470-97305-9 (ebk)

Printed and bound in Great Britain by TJ International, Padstow, Cornwall

10 9 8 7 6 5 4 3 2

WILEY

About the Authors

Nigel Denby trained as a dietitian at Glasgow Caledonian University, following an established career in the catering industry. He is also a qualified chef and previously owned his own restaurant.

His dietetic career began as a Research Dietitian at the Human Nutrition Research Centre in Newcastle upon Tyne. After a period working as a Community Dietitian, Nigel left the NHS to join Boots Health and Beauty Experience where he led the delivery and training of Nutrition and Weight Management services.

In 2003 Nigel set up his own Nutrition consultancy, delivering a clinical service to Hammersmith and Queen Charlotte's Hospital Women's Health Clinic and the International Eating Disorders Centre in Buckinghamshire, as well as acting as Nutrition Consultant for the Childbase Children's Nursery Group.

Nigel also runs his own private practice in Harley Street, specialising in Weight Management, PMS / Menopause and Irritable Bowel Syndrome.

Nigel works extensively with the media, writing for the *Sunday Telegraph Magazine, Zest, Essentials,* and various other consumer magazines. His work in radio and television includes BBC and ITN news programmes, Channel 4's *Fit Farm, BBC Breakfast,* and *BBC Real Story.*

Nigel's first book, *The GL Diet,* was published by John Blake in January 2005.

Sue Baic is a Senior Lecturer in Nutrition and Public Health in the Department of Exercise, Nutrition and Health Sciences at the University of Bristol. She has a first degree from Bristol University followed by a Master of Science in Human Nutrition from London University. Sue is a Registered Dietitian (RD) with over 20 years' experience in the field of nutrition and health in the NHS and as a freelance consultant. She feels strongly about providing nutrition information to the public that is evidence based, up to date, unbiased, and reliable.

As a member of the British Dietetic Association she has spoken and written for the media on a variety of nutrition related health issues. Sue lives in Bristol and spends much of her spare time running up and down hills in the Cotswolds in an attempt to get fit.

Carol Ann Rinzler is a noted authority on health and nutrition and holds an MA from Columbia University. She writes a weekly nutrition column for *The New York Daily News* and is the author of more than 20 health-related books, including *Controlling Cholesterol For Dummies, Weight Loss Kit For Dummies,* and the highly acclaimed *Estrogen and Breast Cancer: A warning for women.* Carol Ann lives in New York with her husband, wine writer Perry Luntz, and their amiable cat, Kat.

Authors' Acknowlededgments

From ND

Thanks go to my writing partner, Sue, for always spotting my typos and for her meticulous attention to detail!

From SB

My part of this book is dedicated to my mother, Pat, who encouraged me to invest the time and effort to enable me to pursue a career that I love. I would also like to thank John and Rosie for their support while I was writing this book and for laughing at my stories even when they've heard them before. Thanks also go to Nigel for asking me to join him in this venture and providing encouragement and much humour along the way.

We would both like to thank the excellent team at Wiley, especially Rachael Chilvers, Alison Yates, Jo Jones and Nicole Hermitage for their part in bringing this book to fruition.

From CAR

Thanks to my husband, Perry Luntz, a fellow writer who, as always, stayed patient as a saint and even-tempered beyond belief while I was racing pell-mell (and not always pleasantly) to deadline.

Publisher's Acknowledgments

We're proud of this book; please send us your comments through our Dummies online registration form located at www.dummies.com/register/.

Some of the people who helped bring this book to market include the following:

Acquisitions, Editorial, and Media Development

Project Editor: Jo Jones

Commissioning Editor: Nicole Hermitage

Assistant Editor: Ben Kemble

Proofreader: Charlie Wilson

Technical Editor: Brigid McKevith

Production Manager: Daniel Mersey

Publisher: David Palmer

Cover Photo: © Wolfgang Kraus - Fotolia.com

Cartoons: Ed McLachlan

Composition Services

Project Coordinator: Lynsey Stanford

Layout and Graphics: Carl Byers, Lavonne Roberts

Proofreader: Jessica Kramer

Indexer: Ty Koontz

Contents at a Glance

Table of Contents

Part IV: Processed Food 247

Introduction

●●

*O*nce upon a time people simply sat down to dinner, eating because they were hungry or just for the pleasure of it. Nobody said, 'I wonder how much trans fat is in the margarine', or asked whether the bread had a low gly-caemic index. Today, the dinner table can be a battleground between health and pleasure. For many people, the fight to eat what does them good rather than what tastes good has become a lifelong struggle.

This book is designed to end the war between your need for a healthy diet and your equally compelling need for tasty meals. In fact, armed with a little insider knowledge you'll find that what's good for you can also be good to eat – and vice versa.

About This Book

Nutrition For Dummies doesn't aim to send you back to the classroom, sit you down, and make you take notes about what to eat every day from now until you're old and grey. Instead, this book means to give you the information you need to make wise food choices – which always means choices that please the palate as well as the body. Some of what you'll read here is about the basics: the roles of vitamins, minerals, proteins, fats, carbohydrates, and even plain water. You'll also read tips about how to put together a nutritious shopping list and how to include all the foods you enjoy as part of a healthy balanced diet.

For those who know absolutely nothing about nutrition except that it deals with food, this book is a starting point containing sound information you can trust. For those who know more than a little about nutrition, this book is a refresher course to bring you bang up to date on what's happened in the field since you last checked.

Conventions Used in This Book

We use the following conventions throughout the text to make things consistent and easy to understand:

✔ All web addresses appear in `mono font`.

✔ New terms appear in *italic* and are closely followed by an easy-to-understand definition.

✔ **Bold** is used to highlight key words and phrases in a list, as well as the action parts of numbered steps.

✔ Nutritionists commonly use metric terms such as gram (g), milligram (mg), and microgram (mcg) to describe quantities of protein, fat, carbohydrates, vitamins, minerals, and other nutrients.

✔ Nutritionists measure food in 100-gram portions. So 'a portion' in this book means a 100-gram dollop unless we state otherwise.

What You Don't Have to Read

You want to get to the important stuff but you're too pushed for time to read it all? We know that feeling and so we've tried to make it that bit easier for you. Some parts of this book are fun or informative but not necessarily vital to your understanding of nutrition. For example:

✔ **Text in sidebars:** The sidebars are the shaded boxes that appear here and there. They share anecdotes and observations but aren't essential reading.

✔ **Anything with a Technical Stuff icon attached:** This information may be interesting but it's not critical to your understanding of nutrition.

Foolish Assumptions

Every book is written with a particular reader in mind, and this one is no different. As we wrote this book, we made the following basic assumptions about who you are and why you paid out your hard-earned money for an entire volume about nutrition:

✔ You didn't study nutrition at school or university but now you've discovered that you have a better chance of staying healthy if you know how to put together a well-balanced, nutritious diet for you and your family.

✔ You're confused by conflicting advice on vitamins and minerals, not to mention newer dietary issues such as antioxidants and low carb diets. You need a reliable road map through the nutrient maze.

✔ You want basic information, but you don't want to become an expert in nutrition or spend hours digging your way through medical textbooks and journals.

How This Book Is Organised

The following is a brief summary of each part in *Nutrition For Dummies*. You can use this guide as a fast way to check out what you want to read first. We've designed this book so you don't have to start with Chapter 1 and read straight through to the end. You can dive in absolutely anywhere and still come up with tons of tasty information about how food helps your body work.

Part 1: The Basic Facts about Nutrition

Chapter 1 defines nutrition and what we mean by essential nutrients. This chapter also tells you how to read a nutrition study and how to find reliable information on nutrition you can trust. Chapter 2 is a really clear guide to how your digestive system works to transform food and drink into the nutrients you need to sustain a healthy body. Chapter 3 explores the reasons why you eat when you eat, the difference between hunger and appetite, and why you like the foods you like.

Part 11: What You Get from Food

Chapter 4 gives you the facts about protein: where you get it and what it does in your body. Chapter 5 does the same job for dietary fat, and Chapter 6 looks at calories – the energy supply to your body. Chapter 7 explains carbohydrates: sugars, starches, and that indigestible but totally vital substance, dietary fibre. Chapter 8 outlines the risks and, yes, some newly proven benefits of alcohol.

Chapter 9 is about vitamins, the substances in food that control so many vital chemical reactions in your body. Chapter 10 is about minerals, substances that go to build so many of our tissues. Chapter 11 explains phytochemicals, newly identified but very important substances in food. Chapter 12 is about water, the essential liquid that comprises as much as 70 per cent of your body weight. This chapter also describes the functions of electrolytes, special minerals that maintain your fluid balance (the correct amount of water inside and outside your body cells).

Part 111: Healthy Eating

Chapter 13 discusses what makes a healthy diet. This chapter is based on recent recommendations from the top nutritional organisations so you know it's good for you! Chapter 14 shows you how to use a food balance model and read food labels to make wise choices when shopping, cooking, and snacking.

Chapter 15 gives you the information to ensure optimum nutrition whoever you are. It examines the key dietary recommendations appropriate for every life stage. Chapter 16 explores how to eat well when you're out and about – at work in restaurants or on the road. Chapter 17 discusses ways to achieve good nutrition standards in institutions like schools and hospitals

Part IV: Processed Food

Chapter 18 asks and answers this simple question: what is processed food? Chapter 19 shows you how cooking, freezing and canning affect the way food looks and tastes, as well as its nutritional value. We also take a peek inside the world of GM food. Chapter 20 gives you the lowdown on food additives – what do they do and are they safe?

Part V: Food and Health

Chapter 21 explains why people are allergic to certain foods and presents strategies for identifying and avoiding the problem foods. Chapter 22 is about how eating or drinking certain foods and drinks may affect your mood – a hot topic these days with nutrition researchers. Chapter 23 looks at food and medicine. It explores which special diets can be used to treat medical problems (as well as which can't). It also covers how foods can interact with drugs and investigates which medications are the most likely to affect your nutritional status. Chapter 24 looks at dietary supplements: who might need them, how they're regulated, and how to choose the most effective types.

Part VI: The Part of Tens

You can't have a *For Dummies* book without The Part of Tens! This part provides ten great nutritional web site addresses, lists ten superfoods – mouthwatering foods we think have positively magical health benefits – and gives you the bottom line on ten trendy weight-loss diets.

Icons Used in This Book

Icons are a handy *For Dummies* way to catch your attention as you slide your eyes down the page. The icons come in several varieties, each with its own special meaning:

Nutrition is full of stuff that 'everybody knows'. This masked marvel clues you in to the real facts when (as often happens) 'everybody's wrong!'

This little guy looks smart because he's marking the place where you find explanations of the terms used by nutrition experts.

This icon alerts you to clear, concise explanations of technical terms and processes – details that are interesting but not necessarily essential to your understanding of a topic. In other words, skip them if you want, but try a few first.

Check out these snippets of useful information that you might want to bear in mind.

Bull's-eye! This is time and stress saving information that you can use to improve your diet and health.

This is a warning icon, alerting you to nutrition pitfalls such as supplements that may do more damage than good to your health.

Where to Go from Here

Ah, here's the best part. *For Dummies* books aren't linear (a fancy way of saying that they don't proceed from A to B to C . . . and so on). In fact, you can dive right in anywhere, say at L, M, or N, and still make sense of what you're reading because we've made sure each chapter delivers a complete message.

For example, if carbohydrates are your passion, go right to Chapter 7. If you want to know how to understand food labels, skip to Chapter 14. If you've always been fascinated by food additives, your choice is Chapter 20. You can use the Table of Contents to find broad categories of information or the Index to look up more specific things.

If you're not sure where you want to go, you may want to start with Part I. It gives you all the basic info you need to understand nutrition and points to places where you can find more detailed information.

Part I
The Basic Facts about Nutrition

'Our Stanley's appearance & personality have certainly changed since he went on the Mediterranean Diet'

In this part . . .

To use food wisely, you need a firm grasp of the basics. In this part, we look at why good nutrition is important and define what we mean by essential nutrients. We explore ways in which you can tell whether information about nutrition is reliable. We also give you a detailed explanation of digestion (how your body turns food into nutrients). Finally in this section, we explain why you eat when you eat – the realm of hunger and appetite – and why you find certain foods more appetising than others – the world of taste and smell.

Chapter 1

What's Nutrition, Anyway?

*A*s you read this book you'll follow a fantastic journey through the body – a journey that carries food from your plate to your mouth, through your digestive system and into every tissue and cell. Along the way, you'll have an opportunity to see how your organs and digestive systems work. You'll discover why some foods are particularly important to your health. And most importantly you'll find out how to manage your diet so that you can get the biggest return (nutrients) from your investment (food).

Why Nutrition Matters

Technically speaking, *nutrition* is the science of how the body takes in and uses food. All living things need food and water just to stay alive. If you want to live *well*, then you need not only food but *good* food, meaning food with the essential nutrients. Without these nutrients:

✔ Your bones can become brittle (not enough calcium or vitamin D).

✔ Your gums may bleed (not enough vitamin C).

✔ You may feel tired and short of breath (not enough iron).

But optimal nutrition isn't just about avoiding deficiency diseases. We now know that a good diet can help to:

✔ Protect against common health problems such as heart disease, stroke, cancer, and high blood pressure (see Chapters 5 and 23)

✔ Provide enough of the right fuel and fluid for regular physical activity (see Chapters 7 and 12)

✔ Improve your mood and your concentration levels (see Chapter 22)

Understanding how a good diet protects against these health problems requires a familiarity with the language and concepts of nutrition. Knowing some basic chemistry is helpful (don't panic: Chemistry can be easy when you read about it in plain English). A smattering of sociology and psychology is also useful, because although nutrition is mostly about how food sustains your body, it's also about the cultural traditions and individual differences that explain how and why we choose food (see Chapter 3).

Nutrition is about why you eat what you eat and how it affects your health and wellbeing.

You are what you eat

I bet you've heard that before! However, it's worth repeating because the human body really is built from the things it gets from food: water, protein, fat, carbohydrates, vitamins, and minerals. Your diet provides the energy and building blocks you need to construct and maintain every cell and organ in your body. To do this you need a range of nutrients from two different and distinct groups:

✔ **Macronutrients (macro = big):** Energy, protein, fat, carbohydrates, and fibre

✔ **Micronutrients (micro = small):** Vitamins and minerals

Daily requirements for *macronutrients* are always in the order of several grams. For example, an average man needs about 55 grams of protein a day and 24 grams of fibre.

Your daily requirements for *micronutrients* are much smaller. For example, the *reference nutrient intake* (RNI) for vitamin C is measured in milligrams (¹⁄₁,₀₀₀ of a gram), while the RNIs for vitamin D, vitamin B12, and folate are even smaller and are measured in micrograms (¹⁄₁,₀₀₀,₀₀₀ of a gram). You can find out much more about the RNIs, including how they vary for people of different ages, in Chapter 15.

Energy from food

Energy is your power supply. Your body cells burn or metabolise virtually every mouthful of food you eat to give you energy, even when the food doesn't give you many other nutrients. The amount of energy released from food in this way is measured in *kilocalories* (kcal) or in *kilojoules* (kJ). Kilojoules is the standard international (SI) unit for energy and as such is the more scientifically accurate way to express energy. However, most of us are more familiar with food energy expressed as kcals or even more usually as calories. One kilocalorie is equal to one calorie, which is equal to 4.18 kilojoules.

You can read more about metabolism in Chapter 2, and Chapter 6 is your source for information about energy. However, all you need to know for now is that food is the fuel on which your body runs. If you don't eat enough food, you won't get enough energy.

Other nutrients in food

Your body needs other nutrients to build, maintain, and repair tissues. Nutrients also empower cells to send messages back and forth and conduct essential chemical reactions, such as the ones that make it possible for you to move, see, hear, eliminate waste, and do everything else natural to a living body.

NUTRITION SPEAK

Essential nutrients for pot plants and pampered pets

Many organic compounds (substances similar to vitamins) and elements (minerals) are an essential part of the diet for your green or furry friends but not for you, because you make them yourself from the food you eat. Two good examples are the organic compounds choline and myoinositol. *Choline* is an essential nutrient for several species of animals, including dogs, cats, rats, and guinea pigs. It is essential for human beings because it forms part of cell membranes and helps form nerve-endings in the brain, but the human body produces choline on its own. You can get extra choline from milk, eggs, liver, and peanuts. *Myoinositol* is

an essential nutrient for gerbils and rats, but human beings synthesise it naturally and use it in many body processes, such as transmitting signals between cells.

Here are some more nutrients that are essential for animals and/or plants but not for you:

Organic compounds	Elements
Carnitine	Nickel
Myoinositol	Silicon
Taurine (but essential in newborn human infants)	Tin
	Vanadium

What's an essential nutrient?

In nutrition speak; an *essential nutrient* is a very precious thing:

- ✔ **An essential nutrient cannot be manufactured in the body.** You have to get essential nutrients from your diet or from a nutritional supplement.

- ✔ **The lack of an essential nutrient in your diet is often linked to a specific deficiency disease.** For example, people who go without protein for extended periods of time develop the protein-deficiency disease *kwashiorkor.* Those who do not get enough vitamin C develop the vitamin C–deficiency disease *scurvy.* A diet or supplement rich in the essential nutrient cures the deficiency disease, but you need the proper nutrient. In other words, you can't cure a protein deficiency with extra amounts of vitamin C.

- ✔ **Not all nutrients are essential for all species of animals.** For example, vitamin C is only essential for human beings, apes, and guinea pigs. All other animals, including cats, dogs, and horses, can make all the vitamin C they need in the liver just from a type of sugar called glucose.

Essential nutrients for human beings include many well-known vitamins and minerals, along with several *amino acids* (the building blocks of proteins) and some fatty acids. Head to Chapters 4, 5, 9, and 10 for more about these essential nutrients.

Other interesting substances in food

One of the latest tremors in the nutrition world has been caused by phytochemicals. *Phyto* is the Greek word for plants, and *phytochemicals* are simply chemicals from plants. Many vitamins are phytochemicals, such as beta carotene, a deep yellow pigment in fruits and vegetables that your body can convert to a form of vitamin A. *Phytoestrogens*, hormone-like chemicals, grabbed the spotlight when it was suggested that a diet high in *isoflavones* (a type of phytoestrogen found in soya beans) may lower the risk of heart disease and cancers of the breast, ovary, and prostate. To find out more about phytochemicals, including phytoestrogens, check out Chapter 11.

Your nutritional status

Nutritional status is a phrase used to describe the state of your health related to your diet. People with a poor diet do not get all the nutrients they need for optimum health and are *malnourished* (mal = bad).Overweight or obese people can still be malnourished! Malnutrition may arise from:

✔ **A diet that does not provide enough food.** This situation may occur in times of famine, or through voluntary starvation because of an eating disorder, or because something in your life disturbs your appetite, such as illness.

✔ **A diet that, while otherwise adequate, is deficient in a specific nutrient or nutrients, such as vitamin C or iron.**

✔ **A rare metabolic disorder that prevents your body from absorbing or metabolising (processing) specific nutrients, such as protein or carbohydrate.**

✔ **A medical condition that prevents your body from using nutrients.** For example, malabsorption is a side effect of many digestive tract disorders such as coeliac disease or inflammatory bowel disease.

Health professionals have many tools with which to rate your nutritional status. They can:

✔ **Review your medical history to see whether you have any conditions** that may make it hard for you to eat certain foods or problems that interfere with your ability to absorb nutrients.

✔ **Perform a physical examination to look for obvious signs of nutritional deficiency or recent unplanned loss of weight.**

✔ **Carry out blood tests that can identify early signs of malnutrition,** such as the lack of red blood cells that characterises anaemia caused by an iron deficiency.

At every stage of life, the aim of a good diet is to maintain a healthy nutritional status.

Finding Nutrition Facts

Getting reliable information about nutrition can be a daunting challenge. Most of your nutrition information is likely to come from television and radio, newspapers and magazines, books, and the Internet. So how can you tell whether what you hear or read is based on sound evidence?

People you can trust about nutrition

The people who make nutrition news can be scientists, reporters, or simply someone who wandered in off the street with a bizarre new theory. (Apricots cure cancer! Never eat bread and cheese at the same time! Eating vegetable soup makes you lose weight!) The following few groups of people *can* give you sound advice you can trust:

✔ **Registered dietitians** (RDs) are the only qualified health professionals who assess, diagnose and treat diet and nutrition problems at an individual and wider public health level. Uniquely, dietitians use the most up to date public health and scientific research on food, health and disease which they translate into practical guidance to enable people to make appropriate lifestyle and food choices.

In the UK, registered dietitians are the only nutrition professionals to be statutorily regulated and governed by an ethical code, to ensure that they always work to the highest standard .The title 'registered dietitian' is protected by the Health Professions Council (HPC). A person with the letters *RD* after his or her name must be suitably qualified and registered with the HPC as being fit to practise within an agreed ethical code of conduct. All registrants of the HPC must commit to continuing professional development to remain registered and call themselves a dietitian. The HPC publishes its online register at http://hpc-portal.co.uk/online-register, so you can check to see whether a dietitian is registered.

Most people can see a registered dietitian within the NHS after a referral by an NHS GP, doctor, health visitor or other medical staff. You can also self-refer. Consultations with dietitians within the NHS are free.

Alternatively if you want to see a registered dietitian who practises privately, you can search on-line for a dietitian near you at the Freelance Dietitians web site, www.freelancedietitian.org, which is run by the British Dietetic Association.

✔ **Nutritionists** are qualified in the study of and research into nutrition and can often offer sound advice about food and healthy eating. A nutritionist usually has a first degree in nutrition or a related science subject, or may be a professional in another field such as medicine. In the UK, the Association for Nutrition (AfN; www.associationfornutrition.org) is a new professional body for the regulation and registration of nutritionists (including public health nutritionists, exercise nutritionists, and animal nutritionists). Nutritionists on the AfN Register have high ethical and quality standards, founded on evidence-based science.

At present the title 'nutritionist' is not protected. As a result, almost anyone can call himself or herself a nutritionist. You can be sure of the credentials only if you choose a registered nutritionist.

✔ **Health reporters and writers** specialise in providing information about the medical and/or scientific aspects of health and food issues. Like reporters who concentrate on politics or sports, health reporters often gain their expertise through years of investigating their field. Most health writers have the scientific background required to make it possible for them to translate technical information into language that non-scientists can understand. Some health reporters are also trained as dietitians or nutritionists.

Research you can trust

You open your newspaper or turn on the evening news and find out that a group of researchers at an impeccably prestigious scientific organisation has published a study showing that yet another food or drink you enjoy is dangerous to your health. For example:

✔ Drinking coffee puts a strain on your heart.

✔ Food additives cause allergic reactions.

So you throw out the offending food or drink or rearrange your daily routine to avoid the once acceptable item. And then what happens? Two weeks, two months, or two years down the road, a second, equally prestigious group of scientists publishes a second study conclusively proving that the first group got it wrong: In fact, coffee has no adverse effect on your heart and may even protect against diabetes. and only certain additives may cause a problem in *some* sensitive individuals.

What's a body made of?

On average approximately 60 per cent of your weight is water, 20 per cent is body fat (slightly less for a man), and 20 per cent is a combination of mostly protein, plus carbohydrates, minerals, vitamins, and other naturally occurring biochemicals.

An easy way to remember this formula is to think of it as the *60–20–20 rule*.

Based on these percentages, you can reasonably expect that an average 70 kilogram person's body weight consists of about:

✔ 40 kilograms of water

✔ 15 kilograms of body fat

✔ 15 kilograms of a combination of protein (up to about 80 per cent), minerals (up to 15 per cent), carbohydrates (up to 5 per cent), and vitamins (a trace).

The exact proportions vary from person to person.

For example, a young person's body has proportionately more muscle and less fat than an older person's, while a woman's body has proportionately less muscle and more fat than a man's. As a result, more of a man's weight comes from protein and calcium, while more of a woman's weight comes from fat. Protein-packed muscles and mineral-packed bones are denser tissue than fat, so weigh a man and a woman of roughly the same height and size, and the man is likely to be the heavier every time.

Who's right? Nobody seems to know. That leaves you on your own to come up with the answer. Never fear – simply ask a few common-sense questions of any study you read about.

Where was the study published?

Studies published in scientific journals are usually *peer reviewed*. This means that an independent group of scientists has looked in detail at the study before it's published. The scientists will have checked that the study was well designed, how it was carried out, and whether the conclusions are appropriate. One of the quickest ways to find information from these studies is to go to a reputable nutrition-related web site. We've taken on some of the leg work and given you ten such web sites in Chapter 25, but another good source you can search for reliable information is www.scholar.google.com.

Does this study include human beings?

Animal studies can alert researchers to potential links between diet and health, but working with animals alone cannot give you conclusive proof. Different species respond differently to various nutrients. Many foods or drugs that harm a laboratory rat won't harm you or are given in such large doses that you would not be at risk from the amount found in a normal diet.

Are enough people in this study?

Any study must include sufficient numbers of participants to have adequate power to show anything useful or applicable to others. If you don't have enough people in the study – several hundred to many thousand – to establish a pattern, some effects may just have occurred by chance. If you don't include different types of people, which generally means young and old men and women of different ethnic groups, the results may not apply across the board. For example, the original studies linking high blood levels of cholesterol to an increased risk of heart disease and small doses of aspirin to a reduced risk of a second heart attack were done only with men. It wasn't until researchers conducted follow-up studies with women that they were able to say with any certainty that high cholesterol is dangerous and aspirin is protective for women as well as men.

Is there anything in the design of this study that can influence its conclusions?

To establish the links between diet and health you need to be able to measure someone's diet. This is easier said than done. You can do it in a *retrospective* study (by asking the participants in a study what they ate in the past, usually by a food frequency questionnaire). However, because memory isn't always accurate people tend to forget what they ate in the past. As a result, this type of study is considered less accurate than a *prospective* study (one that asks people to record what they actually eat as they go along, usually in a food diary). However, even prospective studies have their flaws because they can only ever provide a snapshot of the diet.

The longer the record of someone's diet, the better the picture you get (for instance, seven days is more accurate than three days). Using weighing scales may be more accurate than estimating portion weights. However, the more accurate the method, the greater the burden on the participants. Studies comparing reported food intake (food records) with biological markers of actual food intake have shown that participants often fail to record everything they eat (under recording), or even alter their diet by eating what they think the researchers want or expect (undereating). New technologies try to minimise theses errors by asking people to take photos of what they've eaten with a camera or their mobile phones. These photos are then sent off to researchers who analyse the information.

Other types of study look at the links between diet and health by randomly assigning people to groups and asking them to eat a certain diet. They will then compare various aspects of their health with a control group on their normal diet. But you still don't know whether the participants really ate the diet to which they were assigned. Other studies, known as feeding studies, actually provide the food for their participants so it's more likely they will follow the diet. Bear in mind that errors of measurement occur in any dietary study, so be aware of the limitations of the method used.

Are the study's conclusions reasonable?

When a study comes up with unexpected results, the conclusions need to be examined very carefully. For example, in 1990 the long-running Nurses' Study at the Harvard School of Public Health in the USA concluded that a high-fat diet increased the risk of colon cancer. However, subsequent analysis of the data showed a link only to diets high in red and processed meats. It didn't find any link to diets high in fat from dairy foods. Researchers are still working out whether this finding is really true and whether something other than fat in meat is involved. Later findings from the same nurses' study literally went against the grain. Contrary to prevailing medical wisdom, the results suggested that eating dietary fibre doesn't protect against the risk of colon cancer. Many view these findings with a healthy degree of scepticism, but we're still waiting (over ten years on) for a more definitive answer.

Chapter 2

Digestion: The 24-Hour Food Factory

*W*hen you see or smell something appetising such as freshly baked bread or a plate of roast dinner, your digestive system leaps into action. Your mouth waters. Your stomach rumbles. Intestinal glands begin to secrete the chemicals that turn food into the nutrients that build new tissues and provide the energy you need to keep you ticking.

This chapter introduces you to your digestive system and explains exactly how your body digests the many different kinds of foods you eat, all the while extracting the nutrients you need to keep you alive and well.

Introducing the Digestive System

Digestion is a major performance requiring not a cast of thousands, but a group of digestive organs, each designed specifically to perform a cameo role in the digestion process Your digestive system may never actually win an Oscar, but it certainly deserves a Best Director award for its ability to translate complex food into basic nutrients.

The digestive organs

Although exceedingly well organised, your digestive system is basically one long tube that starts at your mouth, continues down through your throat to your stomach, then on to your small and large intestines, and past the rectum to end at your anus.

In between, with the help of the liver, pancreas, and gall bladder, the digestible parts of everything you eat are converted to simple chemicals that your body can easily absorb to burn for energy or build new tissue. The indigestible residue is bundled off and eliminated at the other end as waste.

Figure 2-1 shows the body parts and organs that comprise your digestive system.

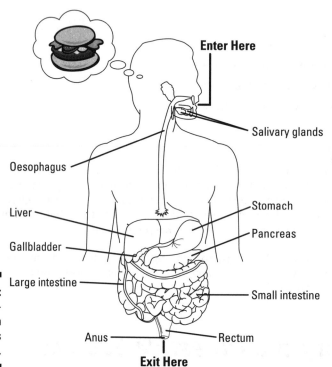

Enter Here

Salivary glands

Oesophagus

Liver

Gallbladder

Large intestine

Stomach

Pancreas

Small intestine

Anus

Rectum

Exit Here

Figure 2-1:
Your digestive system in all its glory.

Digestion: A performance in two acts

Digestion is really a two-part process – half mechanical, half chemical.

✔ *Mechanical digestion* takes place in your mouth and your stomach. Your teeth break food into small pieces that you can swallow without choking. In your stomach, a churning action continues to break food into even smaller particles.

> ✔ *Chemical digestion* occurs at every point in the digestive tract where enzymes and other substances such as *hydrochloric acid* (from cells in the stomach lining) and *bile* (from the gall bladder) dissolve food, releasing the nutrients inside.

Understanding How Your Body Digests Food

Each organ in the digestive system plays a specific role in the digestive drama. But the first act occurs in two places that are never listed as part of the digestive tract: your eyes and your nose.

The eyes and nose

When you see appetising food, you experience a conditioned response. In other words, your thoughts – mmm, that looks good! – stimulate your brain to tell your digestive organs to get ready for action. (Jump to Chapter 3 for more on your conditioned response to food.)

What happens in your nose is purely physical. The aroma of good food is transmitted by molecules that fly from the surface of the food to settle on the membrane lining of your nostrils, stimulating receptor cells on olfactory nerves that stretch from your nose to your brain. When the receptor cells communicate with your brain – something smells delicious! – your brain sends encouraging messages off to your mouth and digestive tract.

The messages say: 'Start the saliva flowing. Warm up the stomach. Alert the small intestine.' In other words, the sight and scent of food make your mouth water and your stomach rumble in anticipatory hunger pangs.

But wait! Suppose you hate what you see or smell? For some people, even the thought of liver is enough to make them nauseous. At that point, your body takes up arms to protect you: you experience a *rejection reaction* – a reaction similar to that exhibited by babies given something that tastes bitter or sour. Your mouth purses and your nose wrinkles as if to keep the food (and its odour) as far away as possible. Your throat tightens and your stomach heaves – muscles contracting not in anticipation but in movements preparatory to vomiting up the unwanted food. Not a pleasant moment.

But for now, we'll assume you like what's on your plate and you take a bite.

The mouth

When you lift your fork to your mouth, your teeth and salivary glands swing into action. Your teeth chew, grinding the food, breaking it into small, manageable pieces. As a result:

- ✔ You can swallow without choking.
- ✔ You start to break up the indigestible layer of fibre surrounding the edible parts of some foods (fruits, vegetables, whole grains) so that your digestive enzymes can get to the nutrients inside.

At the same time, salivary glands under your tongue and in the back of your mouth secrete the watery liquid called *saliva,* which performs two important functions:

- ✔ Moistening and compacting food so that your tongue can push it to the back of your mouth and you can swallow, sending the food down the slide of your throat (oesophagus or gullet) into your stomach.
- ✔ Providing *amylases*, enzymes that start the digestion of complex carbohydrates (starches), breaking the starch molecules into simple sugars. (No protein or fat digestion occurs in your mouth.)

Chewing your food well also helps stimulate the release of digestive juices farther down your gut. It also makes you eat more slowly, giving your brain a chance to recognise when your body has had enough food so helping to stop you overeating.

The stomach

If you were to lay your digestive tract out on a table, most of it would look like a rather narrow tube. The exception is your stomach, a pouchy part just below your throat (oesophagus).

Like most of the digestive tube, your stomach is circled with strong muscles whose rhythmic contractions – called *peristalsis* – move food briskly along and turn your stomach into a sort of food processor that mechanically breaks pieces of food into ever smaller particles. While this is going on, cells in the stomach wall are secreting *stomach juices* – a potent blend of enzymes, hydrochloric acid, and mucus (the mucus protects the stomach from the acid and enzymes). Ugh, it's enough to turn your stomach.

One stomach enzyme – *gastric alcohol dehydrogenase* – digests small amounts of alcohol, an unusual nutrient that can be absorbed directly into your bloodstream even before it's been digested. See Chapter 8 for more about alcohol digestion, including why men can drink more than women without becoming as merry.

Turning starches into sugars

Salivary enzymes don't lay a finger on proteins or fats, but they do begin to digest complex carbohydrates, breaking the long, chainlike molecules of starches into individual units of sugars. You can taste this for yourself with this simple experiment that enables you to experience firsthand the effects of amylases on carbohydrates.

1. **Put a small piece of plain, unsalted cracker on your tongue. No cheese, no pâté, just the cracker, please.**

2. **Close your mouth and let the cracker sit on your tongue for a few minutes.**

   Do you taste a sudden, slight sweetness? That's the salivary enzymes breaking a long, complex starch molecule into its component parts of sugars.

Okay, you can swallow now. The rest of the digestion of the starch takes place farther down, in your small intestine.

Other enzymes, plus stomach juices, begin the digestion of proteins and fats, separating them into their basic components – amino acids (from protein) and fatty acids. (Skip to Chapters 4 and 5 to find out more about amino acids and fatty acids.)

For the most part, digestion of carbohydrates comes to a screeching – though temporary – halt in the stomach, because the stomach juices are so acidic that they inactivate the amylases, the enzymes in your saliva that break complex carbohydrates apart into simple sugars. Stomach acid can break some carbohydrate bonds, so a bit of carbohydrate digestion does take place.

Eventually, your churning stomach blends its contents into a thick soupy mass called *chyme* (from *cheymos*, the Greek word for juice). When a small amount of chyme spills past the stomach into the small intestine, the digestion of carbohydrates resumes in earnest, and your body begins to extract nutrients from food.

The small intestine

Open your hand and put it flat against your belly button, with your thumb pointing up to your waist and your little finger pointing down. Your hand is now covering most of the relatively small space into which your 3 metre (10 foot) long so-called small intestine is neatly coiled. When chyme spills from your stomach into this part of the digestive tube, your body releases a whole new load of digestive juices. These juices include:

- *Alkaline pancreatic juices* that make the chyme less acidic so that amylases (the enzymes that break down carbohydrates) can go back to work transforming complex carbohydrates into simple sugars.

- *Bile*, a greenish liquid (made in the liver and stored in the gall bladder) that enables fats to mix with water.

- *Pancreatic and intestinal enzymes* that finish the digestion of proteins into amino acids (the building blocks for the body) and help digest fat and polysaccharides (type of carbohydrate).

- Bile, a greenish liquid (made in the liver and stored in the gall bladder) that enables fats to mix with water.

Peephole: The first man to watch a living human stomach at work

William Beaumont was a surgeon in the United States Army in the early 19th century. His name survives in the annals of medicine because of an accident that happened on 6 June, 1822. Alexis St Martin, an 18-year-old French Canadian fur trader, was wounded by a musket ball that discharged accidentally, tearing through his back and out of his stomach, leaving a wound that healed but did not close.

St Martin's injury seems not to have affected what must have been a truly sunny disposition: two years later, when all efforts to close the hole in his gut had failed, he granted Beaumont permission to use the wound as the world's first window on a working human digestive system. (To keep food and liquid from spilling out of the small opening, Beaumont kept it covered with a cotton bandage.)

Beaumont's method was simplicity itself. He tied small pieces of food (cooked meat, raw meat, cabbage, and bread) to a silk string, removed the bandage and inserted the string into the hole in St Martin's stomach.

An hour later, he pulled the food out. The cabbage and bread were half digested; the meat, untouched. After another hour, he pulled the string out again. This time, only the raw meat remained untouched, and St Martin, who now had a headache and a queasy stomach, called it a day. But in more than 230 later trials, Beaumont – with the help of his remarkably compliant patient – discovered that although carbohydrates (cabbage and bread) were digested rather quickly, it took up to eight hours for the stomach juices to break down proteins and fats (the meat). Beaumont attributed this to the fact that the cabbage had been cut into small pieces and the bread was porous. Modern nutritionists know that carbohydrates are simply digested faster than proteins and that digesting fats (including those in meat) takes longest of all.

By withdrawing gastric fluid from St Martin's stomach, keeping it at 37.8 degrees centigrade (the temperature recorded on a thermometer stuck into the stomach) and adding a piece of meat, Beaumont was able to time exactly how long it took for the meat to fall apart: ten hours.

Beaumont and St Martin separated in 1833 when the patient, now a sergeant in the United States Army, was posted elsewhere, leaving the doctor to write 'Experiments and Observations on the Gastric Juice and the Physiology of Digestion'. The treatise is now considered a landmark in the understanding of the human digestive system.

While these chemicals are working, peristaltic contractions of the small intestine continue to move the food mass down through the tube so that your body can absorb sugars, amino acids, fatty acids, vitamins, and minerals into cells lining the intestinal wall.

The lining of the small intestine is a series of folds covered with projections like little fingers. The technical name for these small fingers is *villi* (single: villus). Each villus is covered with smaller projections called *microvilli*, and every villus and microvillus is programmed to accept a specific nutrient – and no other. Pretty impressive, eh?

The body absorbs nutrients according to how fast it breaks them down into their basic parts.

- ✔ Carbohydrates, which separate quickly into single sugar units, are absorbed first.

- ✔ Proteins (as amino acids) go next.

- ✔ Fats, which take longest to break apart into their constituent fatty acids, are last.

- ✔ Water-soluble vitamins such as B and C, and minerals are absorbed earlier than those that dissolve in fat.

After you've digested your food and absorbed its nutrients through your small intestine, a number of processes happen:

- ✔ Amino acids, sugars, vitamin C, the B vitamins, minerals including iron, calcium, and magnesium and trace elements are carried through the bloodstream to your liver, where they're processed and sent out to the rest of the body.

- ✔ Fatty acids, cholesterol, and fat soluble vitamins including A, D, E, and K go into the lymph system (another fluid transport system which, like blood, runs throughout the body bathing all the cells). From there they're passed into the blood itself. They, too, end up in the liver, are processed, and are sent out to other body cells.

Inside the cells, nutrients are *metabolised*: burned for heat and energy or used to build new tissues. The metabolic process that gives you energy is called *catabolism* (from *katabole*, the Greek word for casting down). The metabolic process that uses nutrients to build new tissues is called *anabolism* (from *anabole*, the Greek word for raising up).

All aboard the Nutrient Express!

Think of your small intestine as a busy train station – a three-level miniature of Clapham Junction – whose apparent chaos of arrivals and departures is actually an efficient, well-ordered system. (Well, that's what the staff tell you.)

✔ Level 1 is the *duodenum* (at the top, right after your stomach).

✔ Level 2 is the *jejunum* (in the middle).

✔ Level 3 is the *ileum* (the last part before the colon).

This three-level station hums away as nutrients arrive and depart, with millions of 'trains' (the nutrients) running on millions of 'tracks' (the microvilli) designed to accommodate only one kind of train and no other.

The system absorbs and sends on nutrients accounting for more than 90 per cent of all the protein, fat, and carbohydrates that you consume, plus smaller percentages of vitamins and minerals. The train timetable looks something like this:

Level 1   The duodenum   Iron, calcium, magnesium

Level 2   The jejunum   Simple sugars (the end products of carbohydrate digestion) and water-soluble vitamins (vitamin C and the B vitamins other than vitamin B12)

Level 3   The ileum   Amino acids (the end product of protein digestion), fat-soluble vitamins (vitamins A, D, E, and K), fatty acids (the end products of fat digestion), cholesterol, vitamin B12, sodium, potassium, water and alcohol

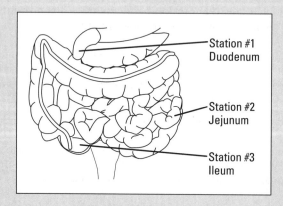

Station #1 Duodenum

Station #2 Jejunum

Station #3 Ileum

The large intestine

After your body has wrung out every useful, digestible ingredient other than water from your food, the rest – indigestible waste – moves into the top of your large intestine, the area known as your *colon*. The colon's primary job is to absorb water from this mixture and then to squeeze the remaining matter into the smelly compact bundle known as faeces.

Faeces (whose brown colour comes from leftover bile pigments) are made of indigestible material from food, plus cells that have sloughed off the intestinal lining, and bacteria – quite a lot of bacteria. In fact, about 30 per cent of the entire weight of the faeces is bacteria. No, these bacteria aren't a sign you're ill. On the contrary, they prove that you're healthy and well. Some of these bacteria are good for you, microorganisms that live in permanent colonies in your colon where they:

✔ Break down previously undigested nitrogen containing compounds, releasing nitrogen gas

✔ Feast on previously indigestible carbohydrates (fibre), producing the gases such as methane, carbon dioxide and hydrogen that sometimes makes you physically uncomfortable (or a social outcast) as well as hydrogen sulphide (a mix of hydrogen with sulphur that makes faeces smell so horrible)

✔ Ferment fibre to produce short chain fatty acids (SCFA) that help protect the cells in the colon against damage from cancer causing chemicals

When the bacteria have finished, the faeces – the small remains of yesterday's food – pass down through your rectum and out through your anus.

Digestion's complete!

Combating Digestive Problems

Eating regular meals based on the principles of healthy eating can help many common digestive complaints. However, specific strategies can also help with digestive problems.

Alleviating common discomforts

Here are three common complaints, and some ideas for easing your discomfort:

✔ **Indigestion** or feeling full and uncomfortable after eating is usually caused by eating too fast or too infrequently, especially eating just one heavy meal late at night. Try to eat smaller meals more often and make time to relax and digest – and enjoy your food!

✔ **Heartburn** (or *gastro oesophageal reflux*) is caused by stomach acid coming backwards into the oesophagus and causing pain, usually after a meal. Drugs such as antacids are very effective treatment but sitting calmly after meals and avoiding bending or lying down can also help. Losing weight from around your waist helps because it reduces abdominal pressure, which makes heartburn worse. Some people also report a benefit from avoiding very spicy or very fatty foods, excessive tea, coffee, tobacco, and alcohol.

✔ **Constipation** is helped by choosing high-fibre foods to stimulate movement of the gut. Try wholegrain bread and breakfast cereals, as well as fruit and vegetables, pulses and beans. Don't forget to drink plenty of fluids to allow the fibre to expand and soften. Try regular physical activity that improves muscle function, including movement of the gut muscle.

Helping to manage irritable bowel syndrome (IBS)

As many as one in ten adults in the UK have IBS and sufferers complain of a range of symptoms varying from severe diarrhoea to constipation, or sometimes both together often with pain, wind and bloating thrown in. Experts still don't really know what causes IBS or how to treat it, or even if what you eat really makes any difference at all. However some people find it helps if they eat regular meals and take time to eat slowly and relax.

Here are some more ideas you can try to relieve IBS symptoms:

✔ **Increase fibre:** When constipation is the main IBS symptom, increasing fibre intake may help. If you try this, do it gradually because any sudden increase may make symptoms worse. Try eating more wholegrain, along with fruit and vegetables, introducing no more than an extra portion every couple of days and make sure you drink plenty of fluid – at least eight cups a day– to help the fibre do its job. Research shows that increasing soluble fibre from oats and linseeds help soften stools and makes them easier to pass. Linseeds may also help if you get symptoms of wind and bloating, and you can add them to foods such as breakfast cereal, yoghurts or soups, up to 1 tablespoon a day.

✔ **Reduce Fibre:** Too much dietary fibre can make pain, bloating, flatulence and diarrhoea worse, so if you suffer any of these you may benefit from reducing fibrous foods such as wholegrain, bran, fruit, vegetable skins as well as pips nuts and seeds. Try limiting fresh and dried fruit to three portions a day and fruit juice to one small glass a day, but remember to make up the recommended 'five-a-day' with vegetables. Avoid sugar-free sweets (such as mints and gum) and food products containing sorbitol. If you have wind and bloating, avoid pulses, onions and brassicas like cabbage.

✔ **Limit resistant starches:** Resistant starches are the starches in foods that the gut doesn't completely digest. Instead, they enter the large bowel, where they're fermented to produce gas. Research has shown that sufferers of IBS may gain some relief from limiting their intake of food high in resistant starches. Try reducing your intake of the following foods which all contain resistant starch:

- Dried pasta – use fresh pasta instead

- Oven chips, crisps, potato waffles, fried rice – choose baked potatoes or boiled rice

- Part-baked and reheated breads, such as garlic bread and pizza base – choose fresh breads

- Processed food such as potato or pasta salad, or manufactured biscuits and cakes

- Pulses, whole grains, sweetcorn, green bananas and muesli that contains bran

- Ready meals containing pasta or potato, such as lasagne, shepherd's pie or macaroni cheese

- Undercooked or reheated potato – instead eat them freshly cooked and still hot

✔ **Keep a food and symptom diary:** This can help you see whether diet affects your symptoms. Remember, symptoms may not be caused by the food you've just eaten, but what you ate earlier in the day or even the day before. And when you try something new to help your IBS, do allow your bowels time to adjust.

✔ **Try probiotics:** Research has found that some people with IBS have changes in the balance of their gut bacteria and that increasing friendly gut bacteria from the probiotics may help (see the following section).

✔ **Check for food allergies/intolerances:** If none of the previous strategies help, you could try a short trial of an exclusion diet with the guidance of a health professional. True food allergy is rare and is seldom the cause of IBS, but sometimes cutting out items such as milk or wheat can help. Make sure you replace the valuable nutrients found in these foods (such as calcium) with an alternative source (see Chapter 10).

For more information on IBS, check out *IBS For Dummies (Wiley),* contact the Gut Trust on 0114 272 3253 or visit their website at www.theguttrust.org.

Sometimes digestive problems can indicate an underlying medical problem so do visit your doctor if any of the following apply to you:

✔ A change in bowel habit to looser and/or more frequent stools for more than six weeks

✔ Digestive problems and a family history of bowel or ovarian cancer

✔ Digestive problems if you're over 60

✔ Rectal bleeding

✔ Unintentional and unexplained weight loss

In these circumstances, before attempting to manage symptoms via your diet, it's important to rule out other medical conditions, and to have a diagnosis established by your doctor or healthcare professional.

Balancing your gut bacteria with probiotics and pre biotics

It may amaze you to know there are ten times more bacteria in your gut than there are cells in your body, and if you gathered them all together they'd weigh around 1 kilogram!

Probiotics are live cultures of the good bacteria normally found in your intestines. They can beneficially affect your health by improving the balance of your gut bacteria, which can be upset by poor diet, stress, ageing, infections or antibiotic use. They help with:

✔ Stimulating your immune system: Probiotics can help you to fight infections better. They can make colds last for less time and have even been found to play a role in preventing the development of eczema in early childhood when taken by the mum during pregnancy.

✔ Preventing diarrhoea: Antibiotics cause major changes to the balance of your gut bacteria, which sometimes results in diarrhoea. Studies show that taking probiotics as soon as you start the antibiotic and continuing for at least one week after the end of the course of antibiotics can help prevent diarrhoea

Probiotics can also reduce the chance of getting traveller's diarrhoea on holiday and some people take them for the entire period they're away.

Probiotics are considered safe for people of all ages unless you have a condition that's harmed your immune system, such as cancer or HIV. They're available as tablets, capsules or sachets or added to a range of foods. The most popular format is as fermented yoghurt style drinks, each of which contains around a billion bacteria. The range and quantity of bacteria is generally much higher than that found in live yoghurt. It's a good idea to experiment before deciding which to buy because brands vary considerably in cost, palatability and effect on different symptoms. You need to take them daily for at least four weeks at the dose recommended by the manufacturer to see whether they're likely to help your symptoms. If a product doesn't work for you try another brand or stop them – the types of bacteria vary between products, so an individual brand won't work for everybody!

Another way of boosting your own natural good gut bacteria is through eating prebiotics. – these are types of fibrous carbohydrate that provide food for the good bacteria in the gut so promoting their growth. Natural sources of prebiotics include onions, garlic, asparagus, artichoke, chicory and bananas. However, in the UK people don't eat large quantities of these and so some people choose to take a prebiotic supplement such as inulin or fructo-oligosaccharide (FOS). You can also find prebiotics added to some probiotic drinks and to some cereal products and bread.

Chapter 3

Why You Eat When You Eat and Like What You Like

*B*ecause you need food to live, your body doesn't hesitate to let you know that it's ready for breakfast, lunch, dinner and probably a few snacks in between. This chapter explains the signals your body uses to get you to the table, and the processes that determine whether you choose the roast beef, the greens or the sticky toffee pudding.

Understanding the Difference between Hunger and Appetite

People eat for two reasons. The first reason is hunger; the second is appetite. Hunger and appetite are *not* the same. In fact, hunger and appetite are entirely different processes.

Hunger is the *need* for food. Hunger is:

✔ A physical reaction that includes chemical changes in your body related to a naturally low level of glucose in your blood several hours after eating

✔ An instinctive, protective mechanism that makes sure your body gets the fuel it requires to function well

Pavlov's performing puppies

Ivan Petrovich Pavlov (1849–1936) was a Russian physiologist who won the Nobel Prize in physiology/medicine in 1904 for his research on the digestive glands. Pavlov's big break, though, was his identification of *respondent conditioning* (also known as a conditioned reflex) – a fancy way of saying that you can train people to respond physically (or emotionally) to an object or stimulus that simply reminds them of something they love or hate.

Pavlov tested respondent conditioning on dogs. He began by ringing a bell each time he offered food to his laboratory dogs so that the dogs learned to associate the sound of the bell with the sight and smell of food. Then he rang the bell without offering the food, and the dogs responded as though food was present – salivating madly, even though the dish was empty.

Food companies are great at using respondent conditioning to encourage you to buy their products: when you see a picture of a deep, dark, rich chocolate bar, doesn't your mouth start to water?

Appetite is the *desire* for food. Appetite is:

- ✔ A sensory or psychological reaction (looks luscious! smells scrummy!) that stimulates an involuntary physiological response (salivation, stomach contractions)
- ✔ A conditioned response to food (see the sidebar on Pavlov's dogs)

The practical difference between hunger and appetite is this: when you're hungry, you eat one handful of peanuts. After that, your appetite may lead you to eat two more handfuls just because they look appealing or taste good.

In other words, appetite is the basis for the familiar saying 'Your eyes are bigger than your stomach'; not to mention the well-known advertising slogan 'Once you pop, you can't stop' – the advertising gurus know exactly how your hunger and appetite work.

Refuelling: The Cycle of Hunger and Satiety

Your body does its best to create cycles of activity that parallel a 24-hour day. Like sleep, hunger occurs at pretty regular intervals, although your lifestyle may make it difficult to follow this natural pattern – even when your stomach loudly announces that it's empty!

Recognising hunger

The clearest signals that your body wants food, right now, are the physical reactions from your stomach and your blood that let you know it's definitely time to eat.

Rumbly tummy

An empty stomach has no manners. If you don't fill it when it's empty, your stomach issues an audible – sometimes embarrassing – call for food. This rumbling signal is called a *hunger pang*.

Hunger pangs are actually plain old muscle contractions. When your stomach's full, these contractions and their continual waves down the entire length of the intestine – known as *peristalsis* – move food through your digestive tract (see Chapter 2 for more about digestion). When your stomach's empty, the contractions squeeze only air, and that makes noise.

Feeling weak and wobbly

Every time you eat, your pancreas secretes *insulin*, a hormone enabling you to move blood sugar (glucose) out of the blood and into cells where it's needed for everyday chores like keeping you breathing and your heart pumping, and enabling you to carry out day-to-day physical work. *Glucose* is the basic fuel your body uses for energy (see Chapter 7). The level of glucose circulating in your blood rises and declines naturally, producing a vague feeling of emptiness, and perhaps weakness, that prompts you to eat. Most people experience the natural rise and fall of glucose as a relatively smooth pattern that lasts about four hours.

Knowing when you're full

The satisfying feeling of fullness after eating is called *satiation*, the signal that says: I can't manage pudding, I've had plenty and I need to leave the table.

Your *hypothalamus*, a small gland in the middle and towards the back of the brain on top of the *brain stem* (the part of the brain that connects to the top of the spinal cord), houses your appetite controls in an area of the brain where hormones and other chemicals that control hunger and appetite are made. The hypothalamus controls the release of neuropeptide Y (NPY) and peptide YY in the gut, two chemicals that tell you when you're full, or if you're empty it sends out a signal: more food!

Other body cells also play a role in knowing when you're full. In 1995, researchers at Rockefeller University in the United States discovered a gene in *fat cells* (the body cells where fat is stored) that directs the production of a hormone called *leptin* (from the Greek word for *thin*). Leptin appears to tell your body how much fat you have stored, thus regulating your hunger. Leptin also reduces the hypothalamus's secretion of NPY, the hormone that signals hunger. When the Rockefeller scientists injected leptin into specially bred fat mice, the mice ate less, burned food faster and lost significant amounts of weight.

Eventually, researchers hope that this kind of information will lead to the creation of safe and effective drugs to combat obesity.

Beating those between-meal energy lows

Throughout the world, the cycle of hunger (namely, of glucose rising and falling) prompts a feeding schedule that generally provides four meals during the day: breakfast, lunch, tea (officially a mid-afternoon meal) and supper or dinner.

Our three-meal-a-day culture forces us to fight our natural rhythm by going without food from lunch at around 1 p.m. to dinner at around 7 p.m. or later. The result is that when glucose levels decline at around 4 p.m., you can get irritable and hungry, and then try to satisfy your natural hunger by grabbing the nearest food, often a high-fat, high-calorie snack.

Some people find that eating six small meals a day suits them much better than three large meals. Good evidence shows that avoiding those 'grab a chocolate bar', low-energy moments mid-morning or mid-afternoon is helpful for weight control.

Maintaining a healthy appetite

The best way to deal with hunger and appetite is to find out how to recognise and follow your body's natural cues.

If you're hungry, eat – in reasonable amounts that support a realistic weight. (Confused about how much you should weigh? Check out the weight table in Chapter 6.) And remember: nobody's perfect. Make one day's indulgence guilt free by reducing your calorie intake and/or increasing your calorie expenditure by exercising proportionately over the next few days. A little give here, a little take there, and you'll stay on target overall.

The key to being a healthy eater – whether you like a few bigger meals or little and often – is balancing out the food you choose. Reaching for nutrient-dense foods rather than energy-dense treats helps keep your weight healthy and also helps to make sure you get all the nutrients you need. Listen to your body, eat what feels right and know when you've had enough and you should do just fine.

Responding to Your Environment on a Gut Level

Your physical and psychological environments definitely affect appetite and hunger, sometimes leading you to eat more or less than normal.

In the bleak mid-winter

You're more likely to feel hungry in a cold environment than a warm place. And you're more likely to want high-calorie dishes in cold weather than in hot weather. Just think about the foods that tempt you in winter – stews, roasts, thick soups – versus those you find tempting on a simmering summer day – salads, chilled fruit, simple sandwiches.

This difference is no accident. Food gives you calories (energy). Calories keep you warm. Making sure that you get what you need, your body processes food faster when it's cold. Your stomach empties more quickly as food speeds along through the digestive tract, which means that those old hunger pangs show up sooner than expected, which, in turn, means that you eat more and stay warmer and . . . well, you get the picture.

Exercising more than your mouth

Everybody knows that exercising gives you a big appetite, yes? Well, actually, no.

People who exercise regularly are likely to have a healthy (read: normal) appetite, but they're rarely hungry immediately after exercising because:

- ✔ Ordinary short bursts of exercise release stored energy (glucose and fat) from your body tissues, so your glucose levels stay steady and you don't feel hungry.

- ✔ Endurance exercise like marathon running or triathalon events eventually use up all stored energy in body tissues. If endurance athletes don't top up with glucose drinks they eventually 'hit the wall' – where the mind wants to keep going, but the body says, 'No, I need more energy!'

- ✔ Exercise slows the passage of food through the digestive tract. Your stomach empties more slowly and you feel fuller for longer.

    Don't eat a heavy meal just before exercising or you may develop cramps or heartburn. Ouch.

- ✔ Exercise (including mental exertion) reduces anxiety. For some comfort-eaters, that means less desire to reach for a snack. Chapter 22 has more on food and mood.

When appetite goes haywire: Eating disorders

An eating disorder is a psychological illness that leads you to eat either too much or too little. Indulging in a hot fudge sundae once in a while isn't an eating disorder. Neither is dieting for three weeks so that you can fit into last year's swimming costume on this year's holiday.

Eating disorders are potentially life-threatening illnesses that require immediate medical attention. The most common eating disorders are anorexia nervosa (self-imposed starvation) and bulimia nervosa (binge eating followed by purging, usually by vomiting).

Treating people with eating disorders is a very specialist field and requires input from a range of health professionals. The Eating Disorders Association (www.b-eat.co.uk) offers information on where to get help and support for sufferers and their families.

Nursing your appetite back to health

Severe physical stress or trauma – a broken bone, surgery, a burn, a high fever – reduces appetite and slows the natural contractions of the intestinal tract. If you eat at times like this, the food can back up in your gut or even stretch your bowel enough to tear it. In severe situations, intravenous feeding – fluids with nutrients sent through a needle directly into a vein – gives you nutrition without danger.

Taking medicine can also affect your appetite and cause you to eat more or less than usual. (See Chapter 23 on food and medicine.)

The fact that a certain medicine affects appetite is almost never a reason to avoid using it. But knowing that a relationship exists between the drug and your desire for food can be helpful.

Tackling Taste: How Your Brain and Tongue Work Together

Your *taste buds* are sensory organs that enable you to perceive different flavours in food – in other words, to taste the food you eat.

Taste buds (also referred to as *taste papillae*) are tiny bumps on the surface of your tongue. Each taste bud contains groups of receptor cells that anchor an antenna-like structure called a *microvillus* that projects up through a pore in the centre of the taste bud. Imagine a thread sticking through the hole in a polo mint. (Chapter 2 covers the microvilli in your digestive tract.)

The microvilli in your taste buds transmit messages from flavour chemicals in the food along nerve fibres to your brain. Your brain translates the messages into perceptions: 'Oh, wow, that's delicious' or 'Ugh, that's revolting'.

The four (or five) basic flavours

Your taste buds definitely recognise four basic flavours: sweet, sour, bitter and salt. Some people add a fifth basic flavour to this list. It's called *umami*, a Japanese word describing richness or a savoury flavour associated with certain amino acids such as glutamates – we talk more about monosodium glutamate (MSG) later in this section – and soya products such as tofu.

Early on, scientists thought that everyone had specific taste buds for specific flavours: sweet taste buds for sweets, sour taste buds for sour, and so on. However, the prevailing theory today is that groups of taste buds work together so that flavour chemicals in food link up with chemical bonds in taste buds to create patterns that you recognise as sweet, sour, bitter and salt. The technical term for this process is *across-fibre pattern theory of gustatory coding*. Try saying that with a mouth full of tofu. Receptor patterns for the main four (sweet, sour, bitter, salt) have been tentatively identified, but the pattern for umami remains elusive.

Flavours aren't frivolous. They're one of the factors that enable you to enjoy food. In fact, flavours are so important that some people use MSG to make food taste better. MSG, most often found in food prepared in Chinese restaurants, stimulates brain cells. People who are sensitive to MSG may actually develop *Chinese restaurant syndrome*, characterised by tight facial muscles, headache, nausea and sweating caused by overactive brain cells. The compound is banned from baby food. However, no real evidence indicates that a little MSG is a problem for people who aren't sensitive to it. That leaves only one question: how does MSG work? Does it enhance existing flavours or simply add that umami flavor on its own? Believe it or not, right now nobody knows. Sorry about that.

Cravings for clay?

Food cravings are a common occurrence during pregnancy. *Pica* is the craving for substances usually considered inedible. Pregnant women crave a wide variety of items, including laundry starch, rocks, matchsticks and clay. No convincing evidence exists to prove that pica has any physiological significance or indicates any mineral deficiency. Thankfully, it's usually a short-lived phenomenon!

The nose knows

Your nose is important to your sense of taste. Just like the taste of food, the aroma of food also stimulates sensory messages. Think about how you sniff your wine before drinking and how the wonderful aroma of baking bread warms the heart and stirs the soul – not to mention the salivary glands. When you can't smell, you can't really taste. As anyone who's ever had a cold knows, when your nose is stuffed and your sense of smell is deadened, almost everything tastes like plain old cotton wool. You can test this theory by closing your eyes, pinching your nostrils shut and having someone put a tiny piece of either a raw onion or a fresh apple into your mouth. Bet you can't tell which is which without looking – or sniffing!

Your health and your taste buds

Some illnesses and medicines alter your ability to taste foods. The result may be partial or total *ageusia* (the medical term for loss of taste). Or you may experience *flavour confusion* – meaning that you mix up flavours, translating sour as bitter, or sweet as salt, or vice versa.

Table 3-1 lists some medical conditions that affect your sense of taste.

Table 3-1	These Things Make Tasting Food Difficult
This Condition	*May Lead to This Problem*
A bacterial or viral infection of the tongue	Secretions that block your taste buds
Injury to your mouth, nose or throat	Damage to the nerves that transmit flavour signals
Radiation therapy to mouth and throat	Damage to the nerves that transmit flavour signals

Tricking your taste buds

Combining foods can short-circuit your taste buds' ability to identify flavours correctly. For example, when you sip wine (even an apparently smooth and silky one), your taste buds say, 'That alcohol's sharp.' Take a bite of cheese first, and the wine tastes smoother (less acidic) because the cheese's fat and protein molecules coat your receptor cells so that acidic wine molecules can't connect.

A similar phenomenon occurs during wine tastings. Try two equally dry, acidic wines, and the second seems mellower because acid molecules from the first one fill up space on the chemical bonds that perceive acidity. Drink a sweet wine after a dry one, and the sweetness is often more pronounced.

Determining Deliciousness

When it comes to deciding what tastes good, all human beings and most animals have four things in common: they like sweets, crave salt, go for the fat and avoid the bitter (at least at first).

These choices are rooted deep in biology and evolution. In fact, you can say that whenever you reach for something that you consider delicious to eat, the entire human race – especially your own ancestors – reaches with you. All right, cavemen didn't have ice cream, but that's what evolution's all about!

Listening to your body

Here's something to chew on: The foods that taste good – sweet foods, salty foods, fatty foods – are essential for a healthy body in the right doses.

✔ Sweet foods are a source of quick energy because your body can convert their sugars quickly to glucose, the molecule that your body burns for energy. (Check out Chapter 6 for an explanation of how your body uses sugars.)

Better yet, sweet foods make you feel good. Eating them tells your brain to release natural painkillers called *endorphins*. Sweet foods may also stimulate an increase in blood levels of *adrenaline*, a hormone secreted by the adrenal glands when circulating glucose is low. The glucose released from sweet foods doesn't stay in the bloodstream for long. It causes a sudden rise in blood glucose when you eat sweet foods but an equally rapid fall shortly after. Adrenaline is sometimes labelled the *fight-or-flight hormone* because it's secreted more heavily when you feel threatened and must decide whether to stand your ground – *fight* – or hurry away – *flight*.

✔ Salt is vital to life. As Chapter 13 explains, salt enables your body to maintain its fluid balance and to regulate chemicals called electrolytes that give your nerve cells the power needed to fire electrical charges that energise your muscles, power up your organs and transmit messages from your brain.

✔ Fatty foods are even richer in calories (energy) than sugars. So the fact that you want them most when you're very hungry comes as no surprise. (Chapters 2 and 5 explain how you use fats for energy.) Which fatty food you want may depend on your sex. Several studies suggest that women like their fats with sugar – where's the chocolate? Men, on the other hand, seem to prefer their fat with salt – bring on the crisps!

Turning up your nose to tastes

Why some people loathe broccoli but love olives (or vice versa) is still something of a mystery to the sensory experts. They suggest, but can't prove, that your food preferences depend on your genes, your culture and your personal experience.

If you're allergic to a food or have a metabolic problem that makes digesting it hard, you may eat the food less frequently but you may enjoy it as much as everyone else does. For example, people who can't digest lactose, the sugar in milk, may end up gassy every time they eat ice cream, but they still like the way the ice cream tastes.

It's important to like your food. The simple act of putting food into your mouth stimulates the flow of saliva and the secretion of enzymes that you need to digest the food. Some studies suggest that if you really like your food, your pancreas releases as much as 30 times its normal amount of digestive enzymes.

Genetics at the dinner table

Virtually everyone instinctively dislikes bitter foods, at least at first tasting. This dislike is a protective mechanism. Bitter foods are often poisonous, so disliking stuff that tastes bitter is a primitive but effective way to eliminate potentially toxic food.

About two-thirds of all human beings carry a gene that makes them more sensitive than usual to bitter flavours. This gene may have given their ancestors a leg up in surviving their evolutionary food trials because they avoided eating poisonous food.

People with this gene can taste very small concentrations of a chemical called phenylthiocarbamide (PTC). Because PTC is potentially toxic, the researchers test for the trait by having people taste a piece of paper impregnated with 6-N-propylthiouracil, a thyroid medication whose flavour and chemical structure are similar to PTC. People who say that the paper tastes bitter are called *PTC tasters*. People who taste only paper are called *PTC nontasters*.

If you're a PTC taster, you're likely to find the taste of saccharin, caffeine, the salt substitute potassium chloride and the food preservatives sodium benzoate and potassium benzoate really nasty. The same is true for the flavour chemicals common to cruciferous vegetables – members of the mustard family, including radishes, cabbage, Brussels sprouts, cauliflower and broccoli.

However, if you truly loathe what you're eating, your body can refuse to take it in. No saliva flows; your mouth becomes so dry that you find it hard to swallow the food. If you do manage to choke it down, your stomach muscles and your digestive tract may convulse in an effort to be rid of the awful stuff.

No such ambivalence exists among people who've vomited after eating a specific food. When that happens to you, you naturally come to like its flavour less. Sometimes, your revulsion may be so strong that you'll never try the food again – even when you know that what actually made you sick was something else entirely, like riding a roller coaster just before eating, or having the flu, or being drunk!

Changing the Menu: Adapting to New and Exotic Foods

New foods are an adventure. As a rule, you may not like them the first time around, but in time – and with patience – what once seemed strange can become just another dinnertime dish.

Exposure to different people and cultures often expands your taste horizons. Some food taboos – horsemeat, snake, dog – may simply be too emotion laden to overcome. Others with no emotional baggage fall to experience. Most people hate very salty, very bitter, very acidic or very slippery foods such as caviar, coffee, Scotch whisky and oysters on first taste, but many later find that they enjoy these foods.

The world has become a smaller place. People think nothing of jumping on a plane and arriving a day later on the other side of the world. With increased travel come new and unusual foods and cuisines; sometimes you may bring ingredients or styles of cooking home with you and they become a regular feature in your diet. Other people decide to make a completely new life in another country and take their foods and tastes with them – sometimes making a living from their native cuisine in their new home. Visit any town and you find restaurants offering fantastic food from every corner of the world. Eating out has moved on from fish and chip shops and steak houses – although a chip butty is hard to beat!

Part II
What You Get from Food

'Another jogger ..... bones picked clean ..... the Milford-on-Thames Cannibal has struck again, sergeant.'

In this part . . .

Here's the lowdown on things you've heard about since your school days: protein, fat, calories, carbohydrates, alcohol, vitamins, minerals, and water. We also cover phytochemicals – a new group of nutrients with near magical health boosting properties.

This is a *For Dummies* book, so you don't have to read straight through from protein to water to see how things work. You can skip from chapter to chapter, back and forth, side to side. Any way you care to use it, this part will clue you in to the value of the nutrients in food.

Chapter 4

Powerful Protein

*P*rotein is an essential nutrient, vital for the structure of your body and its functions. In fact, the word comes from the Greek *protos*, meaning first. Each of the thousands of proteins in your body plays a key role. To visualise a molecule of protein, imagine a very long chain, like a string of sausages. The whole string is the protein and the sausages are *amino acids*, commonly known as the building blocks of protein. In addition to carbon, hydrogen and oxygen atoms, amino acids contain a nitrogen (amino) group. The *amino group* is essential for manufacturing specialised proteins in your body.

In this chapter you find out more, much more, maybe even more than you ever wanted to know about this molecule, how your body uses the proteins you take in as food and how it makes some special ones you need for a healthy life.

Looking Inside and Out: Where Your Body Puts Protein

The human body is chock-a-block with proteins. Proteins are present in the outer and inner membranes of every living cell.

✔ Your hair, your nails and the outer layers of your skin are made of the protein keratin.

 Keratin is a *scleroprotein*, a protein resistant to digestive enzymes. So if you bite your nails, you can't digest them.

✔ Muscle tissue contains myosin, actin, myoglobin and a number of other proteins.

✔ Bone has plenty of protein. The outer part of bone is hardened with minerals such as calcium, but the basic, rubbery inner structure is protein; and bone marrow, the soft material inside the bone, also contains protein.

✔ Red blood cells contain *haemoglobin*, a protein compound that carries oxygen throughout the body. *Plasma*, the clear fluid in blood, contains fat and protein particles known as *lipoproteins*, which ferry cholesterol around and out of the body via the faeces.

Putting Protein to Work: How Your Body Uses Protein

Your body uses proteins to build new cells, maintain tissues and put together new proteins that make it possible for you to perform basic bodily functions.

About half the dietary protein that you consume each day goes into making *enzymes*, the specialised worker proteins that do specific jobs such as digesting food and assembling or dividing molecules to make new cells and chemical substances. To perform these functions, enzymes often need specific vitamins and minerals.

Your ability to see, think, hear and move – in fact, to do just about everything that you consider part of a healthy life – requires your nerve cells to send messages backwards and forwards to each other and to other specialised kinds of cells such as muscle cells. Sending these messages requires chemicals called *neurotransmitters*. Making neurotransmitters requires – surprise! – proteins. *Antibodies*, substances produced in the blood that trigger a reaction when you come into contact with an allergen as well as protecting you from all manner of foreign invaders like viruses, are also made from protein.

Finally, proteins play an important part in the creation of every new cell and every new individual. *Nucleoproteins* are substances made of amino acids and nucleic acids that are present in every cell of your body. See the 'DNA/RNA' sidebar in this chapter for more information about nucleoproteins.

Piling In the Protein: What Happens to the Proteins You Eat

The cells in your digestive tract can absorb only single amino acids or very small chains of two or three amino acids called *peptides*. So digestive enzymes – which are themselves specialised proteins – break proteins from

food into their component amino acids. Then other enzymes
body cells synthesise new proteins by reassembling amino a
compounds that your body needs to function. This process
synthesis. During protein synthesis:

- ✔ Amino acids join up with fats to form *lipoproteins*, the molecules that ferry cholesterol around and out of the body. Or amino acids may join up with carbohydrates to form the *glycoproteins* found in the mucus secreted by the digestive tract.

- ✔ Proteins combine with phosphoric acid to produce *phosphoproteins* such as *casein*, a protein in milk. Phosphorus and calcium (also found in milk) are closely linked in many of their functions and so phosphoproteins make a very handy package.

- ✔ Nucleic acids combine with proteins to create nucleoproteins, essential components of the cell nucleus, and *protoplasm*, the living material inside each cell.

Your body converts the carbon, hydrogen and oxygen that are left over after protein synthesis to glucose, used for energy (see Chapter 7). The nitrogen residue (ammonia) is processed by the liver, which converts the ammonia to urea. Most of the urea produced in the liver is excreted through the kidneys in urine; very small amounts are taken up by skin, hair and nails.

Every day you reuse more proteins than you get from the food you eat, so you need a continuous supply to maintain your protein levels. If your diet doesn't contain sufficient amounts of proteins, you start digesting the proteins in your body, including the proteins in your muscles and, in extreme cases of starvation, your heart muscle.

Examining Protein Types: Not All Proteins Are Created Equal

All proteins are made of building blocks called amino acids, but not all proteins contain all the amino acids you require. However, a varied diet can go a long way to providing the most useful proteins for your body.

Essential and non essential proteins

To make all the proteins that your body needs, you require 21 different amino acids. Ten are *essential amino acids*, which you must obtain from food. The rest are *nonessential* amino acids that you can manufacture yourself from fats, carbohydrates and other amino acids if you don't get them in food.

The essential amino acids are	*The nonessential amino acids are*
Histidine	Alanine
Isoleucine	Arginine (essential in infants)
Leucine	Aspartic acid
Lysine	Asparagine
Methionine	Cysteine (semi essential can be made from methionine)
Phenylalanine	
Selenocysteine	Glutamic acid
Threonine	Glycine
Tryptophan	Hydroxyproline
Valine	Proline
	Serine
	Tyrosine (semi essential can be made from phenylalanine)

TECHNICAL STUFF

DNA/RNA

Would you believe it if we said that sugar and protein determine what colour your eyes are, whether your hair is straight or curly and even what kind of sport you have a natural talent for? No? Then read on. This stuff is quite technical but the bottom line is that every characteristic that makes each and every one of us unique starts off as a sugar or a protein compound.

Nucleoproteins are chemicals in the nucleus of every living cell. They're made of proteins linked to *nucleic acids* – complex compounds that contain phosphoric acid, a sugar molecule, and nitrogen-containing molecules made from amino acids.

Nucleic acids are molecules found in the chromosomes and other structures in the centre of your cells. Nucleic acids carry the genetic codes – genes and chromosomes that determine who you are. They contain one of two sugars, either *ribose* or *deoxyribose*. The nucleic acid containing ribose is called *ribonucleic acid* (RNA). The nucleic acid containing deoxyribose is called *deoxyribonucleic acid* (DNA).

DNA, a long molecule with two strands twisting about each other (the double helix), carries and transmits the genetic inheritance in your chromosomes. In other words, DNA supplies instructions that determine how your body cells are formed and how they behave. RNA, a single-strand molecule, is created in the cell according to the pattern determined by the DNA. Then RNA carries DNA instructions to each cell.

DNA is the most distinctly unique thing about your body. The chances that another person on Earth has exactly the same DNA as you are really small. That's why DNA analysis is used increasingly in identifying criminals or exonerating the innocent. Some people are even proposing that parents store a sample of their children's DNA so that they'll have a conclusive way of identifying a missing child, even years later.

High-quality and low-quality proteins

Because an animal's body is similar to a human's, its proteins contain similar combinations of amino acids. That's why nutritionists call proteins from foods of animal origin – meat, fish, poultry, eggs and dairy products – *high-quality proteins*. Your body absorbs these proteins really efficiently and can use the proteins without much waste to synthesise other proteins. The proteins from plants – grains, fruit, vegetables, legumes (nutrition-speak for peas or beans), nuts and seeds – often have limited amounts of one or more essential amino acids, meaning that their nutritional content isn't as high as animal proteins. The prime exception is the soya bean, a legume packed with abundant amounts of all nine essential amino acids. Soya beans are an excellent source of proteins for vegetarians, especially vegans (people who avoid all products of animal origin, including milk and eggs).

Nowadays most of the world's major health organisations assess the relative protein quality of different foods using a complex formula which considers their essential amino acid composition and their digestibility in the human body to give an overall amino acid score. See Table 4-1 for the relative protein quality of a selection of foods:

Table 4-1 Scoring the Realtive Protein Quality of Different Foods

Food	Amino acid score
Egg	100
Milk (cow's whole)	100
Beef	92
Soya beans	91
Chick peas	78
Lentils	52
Peanuts	52
Wheat	42

Source: FAO/WHO [1990]. Expert consultation on protein quality evaluation. Food and Agriculture Organization of the United Nations, Rome.

Homocysteine and your heart

Homocysteine is an *intermediate,* a chemical released when you digest protein. Unlike other amino acids, which are vital to your health, homocysteine can be hazardous to your heart, raising your risk of heart disease by attacking cells in the lining of your arteries by making them reproduce faster (the extra cells may block your coronary arteries) or by causing your blood to clot, which can also block your arteries.

Years and years ago, before cholesterol moved into the limelight, some clever heart researchers labeled homocysteine the major nutritional culprit in heart disease. Today, they've been vindicated. The British Heart Foundation now lists high homocysteine levels as an independent risk factor for heart disease, perhaps explaining why some people with low cholesterol have heart attacks.

But wait! The good news is that information from several international studies suggests that a diet rich in the B vitamin folate lowers blood levels of homocysteine. Most fruits and vegetables provide plentiful amounts of folate and particularly good sources are spinach, asparagus and cabbage. Eating your five portions a day may protect your heart.

Complete proteins and incomplete proteins

Another way to describe the quality of proteins is to say that they are either complete or incomplete, or have a high or low biological value. A *complete protein* with a *high biological value* is one that contains ample amounts of all essential amino acids; an *incomplete protein* or one with a *low biological value* doesn't. A protein low in one specific amino acid is called a *limiting protein* because it can build only as much tissue as the smallest amount of the necessary amino acid. You can improve the quality in a food containing incomplete or limited proteins by eating it with protein that contains sufficient amounts of amino acids. Matching foods to create complete proteins is called *complementarity.*

For example, grains are low in the essential amino acid lysine, and beans are low in the essential amino acid methionine. By combining the two by eating beans on toast, you improve (or complete) the proteins in both. Another example is combining grains with dairy products, such as eating pasta and cheese. Pasta is low in the essential amino acids lysine and isoleucine, but milk products (like cheese) have abundant amounts of these two amino acids. Shaking Parmesan cheese onto pasta creates a higher-quality protein dish. In each case, the foods have complementary amino acids.

Other examples of complementary protein dishes are peanut butter with bread, and milk with cereal. Many combinations of two or more types of plant proteins are a natural and customary part of the diet in parts of the world where animal proteins are scarce or very expensive. Table 4-2 shows categories of foods with incomplete proteins. Table 4-3 shows how to combine foods to improve the quality of their proteins. Improving proteins is particularly important for people following a vegan or vegetarian diet.

Table 4-2	Foods with Incomplete (Limited) Proteins
Category	*Examples*
Grain foods	Barley, wheat and wheat flour products (such as bread, pasta and couscous), bulgar wheat, maize or corn, rice, rye, oats
Legumes	Black beans, black-eyed peas, cannelloni beans, butter beans, kidney beans, lima beans, lentils, peanuts (and peanut butter), peas, split peas, chick peas
Nuts and seeds	Almonds, Brazil nuts, cashews, pecans, walnuts, pumpkin seeds, sesame seeds (tahini), sunflower seeds

Table 4-3	How to Combine Foods to Complement Proteins	
This Food	*Complements This Food*	*Examples*
Whole grains	Legumes (beans)	Rice and peas; beans on toast; dahl and rice; tofu and noodles; tortilla and refried beans; couscous and chick peas; falafel and pitta bread
Dairy products	Whole grains	Cheese sandwich with wholemeal bread; wholemeal pasta with cheese; pancakes (wheat and milk and egg batter); porridge and milk
Legumes (beans)	Nuts and seeds	Mixed bean salad with sesame seed dressing; hummus
Dairy products	Legumes (beans)	Baked beans with grated cheese
Dairy products	Nuts and seeds	Yogurt with chopped nuts; muesli with milk

The lowdown on gelatin and your fingernails

Some people believe that gelatin is a protein that strengthens fingernails. Sorry, it's just an old wives' tale! Gelatin is produced by treating animal bones with acid, a process that destroys the essential amino acid *tryptophan*, actually reducing the quality of the protein. Scoffing cube after cube of jelly in the hope of growing the perfect set of back scratchers is really a waste of time and jelly!

Deciding How Much Protein You Need

The Department of Health, which sets the requirements (known as *Dietary Reference Values*) for vitamins and minerals, also sets goals for daily protein consumption. As with other nutrients, the department has different recommendations for different groups of people; young or older, men or women.

Calculating the correct amount

As a general rule, the Department of Health says that healthy people need to get around 15 per cent of their daily calories from protein. The reference nutrient intake (RNI) of protein for an adult woman is 45grams and for a man it is 55.5 grams. You can easily obtain these amounts from two average servings of lean meat, fish or poultry (each serving will give around 20 grams of protein) plus two to three servings of dairy foods (gives another 12-18 grams of protein). If you are a vegetarian, you could get your protein from two eggs for example (each gives 6 grams of protein) plus 3 servings of dairy or soya (another 18 grams), and five or six servings of starchy foods such as bread, rice , potatoes or cereal (another 12 grams of protein). The number crunchers amongst you will quickly spot that these combinations quickly add up to the RNIs. The point we're making is that getting all the protein you need from a normal healthy diet is very easy, so you can mix and match your protein sources without worrying whether you're getting enough.

As you grow older you synthesise new proteins less efficiently, so your muscle mass (protein tissue) diminishes while your fat content stays the same or rises. This change is why some people erroneously believe that muscle 'turns to fat' in old age, and where the term 'middle-age spread' comes from (and you thought it was a new type of margarine?). Of course, you still use protein to build new tissue, including hair, skin and nails, which continue to grow until you die. (By the way, the idea that nails carry on growing after death – think corpse scenes from old 'Hammer House' horror movies – arises from the fact that after death, tissue around the nails shrinks, making a corpse's nails simply look longer.) This doesn't mean older people need to suddenly start taking protein supplements, but it does mean that making sure you eat two good quality sources of protein every day is a good idea.

Super soya: The special protein food

Nutrition fact no. 1: Food from animals has complete proteins. **Nutrition fact no. 2:** Vegetables, fruits and grains have incomplete proteins. **Nutrition fact no. 3:** Nobody told the soya bean.

Soya beans have complete proteins with sufficient amounts of all the amino acids essential to human health. In fact, food experts rank soya proteins nearly on a par with egg whites and casein (the protein in milk), the two proteins your body finds easiest to absorb and use (see Table 4-1).

Some nutritionists think that soya proteins are even better than the proteins in eggs and milk, because the proteins in soya have very little of the saturated fat known to clog your arteries and raise your risk of heart attack. Better yet, more than 20 recent studies suggest that eating soya protein as part of a low-fat diet can actually lower your cholesterol levels.

A serving of 113 grams of cooked soya beans has 14 grams of protein; the same amount of tofu has 13 grams. Either serving gives you approximately twice the protein you get from one large egg or one large glass of skimmed milk, and two-thirds of the protein in the same amount of lean minced beef. A large glass of low fat soya milk has 7 grams of protein, a mere 1 gram less than a similar serving of skimmed milk, Soya beans are also friendly to your digestive tract because they're jam-packed with soluble dietary fibre, which helps move food through your digestive tract.

People don't often eat soya beans as nature intended. As they stand soya beans can bepretty unpalatable, which is why those clever food manufacturers turn the beans into drinks, spreads, cheese, meat substitutes, yoghurts and desserts, amongst other things. However more recently palatable forms of fresh or frozen soya bean or edamame (baby soya beans) have started to become more easily available and make a delicious addition to a soup, salad or casserole. Dietitians usually recommend three servings of soya foods, giving around 26 grams of soya protein a day, as a good guide for intake.

If you choose to use more soya (or any other beans), take it slowly – a little today, a little more tomorrow and a little bit more the day after that. Adding too much fibre to your diet in one go can have dramatic effects in the bathroom and lead to pretty unsociable gas, so take it steady and make sure that you increase your fluid intake at the same time.

Dodging protein deficiency

The first sign of protein deficiency is likely to be muscle weakness, because the body tissues are most reliant on protein. For example, children who don't get enough protein have stunted growth and shrunken, weak muscles. They may also have thin hair and sores on their skin, and blood tests may show that the level of albumin in their blood is below normal. *Albumin* is a protein that helps maintain the body's fluid balance and is often measured in blood tests to assess your protein status.

A protein deficiency may also show in other ways in your blood. Red blood cells live for only 120 days. Your body needs protein to produce the haemoglobin for new red blood cells. People who don't get enough protein may become *anaemic*, having less haemoglobin to carry oxygen than they need. Protein deficiency may also show up as fluid retention (*oedema*), an increased risk of infection, hair loss and muscle wasting caused by the body's attempt to get energy by digesting the proteins in its own muscle tissue as an alternative source of calories. That's why victims of starvation are, literally, skin and bones.

Given the high protein content of a normal UK diet (which generally provides far more protein than you actually require), protein deficiency is rare. However, protein deficiency can occur in patients who've undergone burns surgery (see the following section), suffered other forms of injury or medical trauma or as a consequence of an extreme eating disorder such as *anorexia nervosa* (refusal to eat).

Boosting your protein intake: Special considerations

Anyone who's building new tissue quickly needs more than 0.75 grams of protein per kilogram (2.2 pounds) of body weight per day. For example:

- ✔ Infants need as much as 2.0 grams of protein for every kilogram of body weight per day.

- ✔ Adolescents need as much as 1.2 grams per kilogram per day.

- ✔ Pregnant women need an extra 10 grams a day, and breastfeeding women also need extra protein. This extra protein is used to build the foetal tissues and then to produce adequate amounts of nutritious breast milk.

Injuries also raise your protein requirements. An injured body releases above-normal amounts of protein-destroying hormones from the pituitary and adrenal glands. You need extra protein to protect existing tissues, and after severe blood loss, you need extra protein to make new haemoglobin. Cuts, burns or surgical procedures mean that you need extra protein to make new skin and muscle cells. Fractures mean that you need extra protein to make new bone. The need for protein is so important when you've been badly injured that if you can't take protein by mouth, you'll be given an intravenous solution of amino acids with glucose (sugar) or emulsified fat.

Do athletes need more proteins than the rest of us? Recent research suggests that the answer may be yes, but athletes easily meet their requirements, no more than about an additional 0.5 grams per kilogram of body weight per day – somewhere around 25-30 grams extra per day, within the framework of a normal diet which covers their increased appetite and energy needs. Those protein supplement powders and drinks aren't really necessary.

Avoiding protein overload

You can get too much protein. Too much can put a strain on the kidneys – especially in people with diabetes – and it may also increase calcium losses from the bones. Several medical conditions make it difficult for people to digest and process proteins properly. As a result, waste products build up in different parts of the body.

People with liver disease or kidney disease either don't process protein efficiently into urea or don't excrete it efficiently through urine. The result may be uric acid, kidney stones or *uraemic poisoning* (an excess amount of uric acid in the blood).

In recent years the market has been flooded with protein supplements. Some can be useful in specific cases, but others make wildly inaccurate health claims and in some cases can be dangerous. As always, if you're worried that you may need to supplement your diet, it's best to consult your dietitian.

Chapter 5

The Lowdown on Fat and Cholesterol

. . . .

. . . .

The chemical family name for fats and related compounds such as cholesterol is *lipids* (from *lipos*, the Greek word for fat). Liquid fats are called
oils; solid fats are called, well, *fat*. With the exception of *cholesterol* (a fatty
substance that has no calories and provides no energy), fats are high-energy
nutrients. Gram for gram, fats have more than twice as much energy potential
(calories) as protein and carbohydrates (affectionately referred to as carbs):
9 calories per fat gram versus 3.75 calories per gram for proteins and 4 for
carbs. (For more calorie facts, see Chapter 6.)

In this chapter, we cut the fat away from the subject of fats and home in on
the essential facts you need to put together a diet with enough fat (yes, you
do need fat) to provide the bounce that every diet requires. And then we deal
with that ultimate baddie: cholesterol. Surprise! You need some of that too.

Finding the Facts about Fat

Fats are sources of energy that add flavour and texture to food – the sizzle on
the steak, you could say. However, as most people know, fats can also be hazardous to your health. The trick is separating the good fats from the bad and
getting the balance right.

Understanding why your body needs fat

A healthy body needs fat. Your body uses *dietary fat* (the fat that you get from food) to make tissue and manufacture biochemicals, such as hormones. Some of the body fat made from food fat is *visible*. Even though your skin covers it, you can *see* the fat in the *adipose* (fatty) *tissue* in female breasts, hips, thighs, buttocks and belly (all of which are part of the design to make women able to carry, bear and feed children), or the male abdomen and shoulders.

This *visible* body fat:

- Provides a source of stored energy
- Gives shape to your body
- Cushions your skin (imagine sitting in a chair for a while to read this book without your buttocks to cushion your bones)
- Acts as an insulation blanket that reduces heat loss

Other body fat is *invisible*. You can't see this body fat because it's tucked away in and around your internal organs. This hidden fat is:

- Part of every cell membrane (the outer skin that holds each cell together).
- A component of *myelin*, the fatty material that sheathes nerve cells and makes it possible for them to fire the electrical messages that enable you to think, see, speak, move and perform the multitude of tasks natural to a living body. Brain tissue is also rich in fat.
- A shock absorber that protects your organs as much as possible if you fall or injure yourself.
- A constituent of hormones and other biochemicals, such as vitamin D and bile.

Releasing energy from fat

Although fat has more energy (calories) per gram than proteins and carbohydrates, your body has a more difficult time releasing the energy from fatty foods. When you drop a balloon into water, it floats. That's exactly what happens when you swallow fatty foods. The fat floats on top of the watery food and liquid mixture in your stomach, limiting the effect that *lipases* – fat-busting digestive enzymes in the mix below – can have on it. Because fat is digested more slowly than proteins and carbohydrates, you feel fuller (a condition called *satiety*) for longer after eating high-fat food. In fact, very little digestion of fats takes place in the stomach at all – the fatty foods wait until they move through to the small intestine before the action starts.

Into the intestines

When the fat moves down your digestive tract into your small intestine, an intestinal hormone called *cholecystokinin* signals to your gall bladder for the release of bile. *Bile* is an emulsifier, a substance that enables fat to mix with water so that lipases can start breaking the fat into glycerol and fatty acids. These smaller fragments may be stored in special cells (fat cells) in adipose tissue, or they may be absorbed into cells in the intestinal wall where they're either:

- ✔ Combined with oxygen (or burned) to produce heat/energy, plus water and the waste product carbon dioxide, or
- ✔ Used to make lipoproteins that transport fats, including cholesterol, through your bloodstream

Into the body

Glucose, the molecule you get by digesting carbohydrates, is the body's basic source of energy. Burning glucose is easier and more efficient than burning fat, so your body always goes for carbohydrates first. But if you've used up all your available glucose after prolonged exercise or starvation, then it's time to start on your body fat.

The first step in utilising fat as an energy source is for an enzyme in your fat cells to break up triglycerides from your fat stores (fats you burn for energy). The enzyme action releases glycerol and fatty acids, which travel through your blood to body cells where they combine with oxygen to produce heat/ energy, plus water and the waste product carbon dioxide. High-protein/ high-fat/low-carb weight loss diets such as the Atkins regimen means that you burn fat without glucose products from carbohydrates. This can cause weight loss in part because you lose a lot of water. The heavy water losses are simply a result of your body having to use ingested fat for energy rather than its preference for carbohydrate. Burning fat without glucose produces a second waste product called *ketones*. In extreme cases, high concentrations of ketones (a condition known as *ketosis*) alter the acid/alkaline balance (or pH) of your blood and may send you into a coma. Left untreated, ketosis can lead to death. Medically, this condition is mostly seen among people with diabetes. For people on a low-carb diet, the more likely sign of ketosis is smelly urine and breath that smells like acetone (nail polish remover). Yuck!

Focusing on the fats in food

Food contains three kinds of fats: triglycerides, phospholipids and sterols.

Triglycerides are the fats that you use to make adipose tissue and burn for energy.

Phospholipids are hybrids – part lipid, part *phosphate* (a molecule made with the mineral phosphorus) – that act as transporters, ferrying hormones and fat-soluble vitamins A, D, E and K through your blood and backwards and forwards in the watery fluid that flows across cell membranes. By the way, the official name for fluid around cells is *extracellular fluid*. But for us, 'watery fluid' will do nicely!

Sterols are fat and alcohol compounds with no calories. Vitamin D is a sterol. So is the sex hormone testosterone. And so is cholesterol, the base from which your body makes some hormones and vitamins. So you see, not all cholesterol is bad.

Getting the right amount of fat

Getting the right amount of fat in your diet is a delicate balancing act. Too much, and you increase your risk of obesity, diabetes, heart disease and some forms of cancer. (The risk of colon cancer seems to be linked to a diet high in fat.) Too little fat, and infants can't thrive, children don't grow and everyone, regardless of age, is unable to absorb and use fat-soluble vitamins that smooth the skin, protect vision, bolster the immune system and keep reproductive organs functioning.

How much fat do you need in a healthy diet? The dietary reference values (DRVs) for food energy and nutrients for the UK (more about DRVs in Chapter 15) recommend getting no more than 35 per cent of your calories from total fat and no more than 10 per cent from saturated fat. See the section 'Exploring the relationship between fatty acids and dietary fat' in this chapter for more on saturated fat.

The Institute of Grocery Distributors in conjunction with the Department of Health and Ministry of Agriculture, Fisheries and Food developed guideline daily amounts (GDAs) of fat for average adults of normal weight. The guidelines suggest that men should consume up to 95 grams of total fat per day, of which no more than 30 grams should be saturated fat. Women should aim to consume up to 70 grams total fat with no more than 20 grams saturated fat per day. The GDA recommendations for fat assume that men and women consume the recommended number of calories per day, whereas DRVs look at the amount of fat in the diet as a proportion of calories.

To make it simpler when you're shopping, if a label says a food contains more than 20 grams of total fat and more than 5 grams of saturates per 100 grams, it contains a lot of fat. If the food contains less than 3 grams of total fat and 1 gram of saturated fat it contains 'a little' fat.

Your body doesn't need to get either saturated fats, cholesterol or trans fats from food because all of the positive benefits of dietary fats can be met by others like polyunsaturated and monounsaturated fats. (*Trans fats* are fats that are adapted for food manufacturing purposes and appear in some spreads, cakes, biscuits and pastry goods.) The reality is that less desirable fats like saturates, cholesterol and trans fats feature in so many everyday foods and in some cases in foods with other nutritional benefits, so to simply recommend you just don't eat them at all would be unrealistic. DRVs have set maximum levels for the amount these nutrients should contribute to the intake of total fat. So the general message is to try to keep them to a minimum.

Daily recommendations also exist for two essential fatty acids, *alpha-linolenic acid* (also known as an omega n-3) and *linoleic acid* (an omega n-6 fatty acid) – see the nearby sidebar 'Essential fatty acids' for more. Alpha-linolenic acid is found in fish oils and some veggie oils such as soya, rapeseed and linseeds. Linoleic acid is mainly found in nuts and seeds including sunflower, safflower and corn oil.

Table 5-1 shows the DRVs for total fat and the different types of fat, along with the amounts people really eat.

Table 5-1	Daily Fat Intake as a Proportion of Energy for the British Adult Population	
Type of Fat	*Current Average UK Intake*	*Recommended UK Intake*
Total fat	36%	35% or less
Saturated fat	13%	11% or less
Monounsaturated fat	20%	13%
n6 polyunsaturated fat	9%	6.5%
n3 polyunsaturated fat	0.2g	0.2g minimum
Trans fatty acids	1.2%	2% or less

Source: The National Diet & Nutrition Survey: Adults aged 19 to 64 years
L Henderson, J Gregory, G Swan - 2003 - 212.58.231.23

Essential fatty acids

An *essential fatty acid* is one that your body needs but lacks the enzymes to assemble from other fats. You have to get it whole and from food. Linoleic acid, found in vegetable oils, is an essential fatty acid. The other is alpha-linolenic acid. You're unlikely to experience a deficiency of the two fatty acids as long as 2 per cent of the calories you get each day come from fat – and most of us far exceed that!

WARNING!

These recommendations are for adults. Although many organisations such as the British Heart Foundation and the British Dietetic Association recommend restricting fat intake for older children, they stress that infants and toddlers require fatty acids for proper physical growth and mental development. That's why Mother Nature made human breast milk so high in *essential* fatty acids. Never limit the fat in a young child's diet without checking first with your doctor or dietitian.

Finding fat in all kinds of foods

As a general rule:

- Fruits and vegetables have only traces of fat, primarily unsaturated fatty acids.

- Grains have small amounts of fat, up to 3 per cent of their total weight. This is also mainly unsaturated fatty acids.

- Dairy products vary. Cream is a high-fat food. Normal milks and cheeses are moderately high in fat. Skimmed milk and skimmed milk products are low-fat foods. Most of the fat in any dairy product is saturated fatty acids.

- Red meat can be low to moderately high in fat depending on the cut, the breeding and the species. Saturated fatty acids are a large part of the fats in red meat .

- Poultry without the skin is relatively low in fat.

- Fish can be high or low in fat. Fish with darker flesh (such as salmon, trout or mackerel) is usually higher in fat. Its fats are composed primarily of unsaturated fatty acids.

✔ Vegetable oils, butter and lard are high-fat foods. Most of the fatty acids in vegetable oils are unsaturated; with the exception of coconut oil and cocoa butter most of the fatty acids in lard and butter are saturated.

✔ Processed foods, such as cakes, breads and tinned or frozen meat products and vegetable dishes, are generally higher in fat (especially saturated and trans fats) than plain grains, meats, fruits and vegetables.

Here's a simple guide to finding which foods are high (or low) in fat. Oils are virtually 100 per cent fat. Butter and lard are close behind. After that, the fat level drops, from 70 per cent for some nuts down to 2 per cent for most bread. Here's the rule to take away from these numbers: a diet high in grains and plants is always lower in fat than a diet higher in meat and oils. Read food labels to help you decide whether a food is high or low in fat. If a food contains more than 20 grams total fat per 100 grams it's classed as containing 'a lot' of fat. Three grams or less per 100 grams total fat means that the food contains 'a little'. Likewise, more than 5 grams saturated fat per 100 grams is 'a lot' and less than 1 gram saturated fat per 100 grams is 'a little'. Simple!

Defining fatty acids

Fatty acids are the building blocks of fats. Chemically speaking, a *fatty acid* is a chain of carbon atoms with hydrogen atoms attached and a *carbon–oxygen–oxygen–hydrogen group* (the unit that makes it an acid) at one end.

Comparing saturated and unsaturated fatty acids

Nutritionists characterise fatty acids as saturated, monounsaturated or polyunsaturated, depending on how many hydrogen atoms are attached to the carbon atoms in the chain (we describe these categories in detail in the following section). The more hydrogen atoms, the more saturated the fatty acid, and the more likely the fat is to be solid at room temperature.

All the fats in food are combinations of fatty acids.

✔ A *saturated fat*, such as butter, has mostly saturated fatty acids. Saturated fats are solid at room temperature and get harder when chilled.

✔ A *monounsaturated fat*, such as olive oil, has mostly monounsaturated fatty acids. Monounsaturated fats are liquid at room temperature; they get thicker when chilled.

✔ A *polyunsaturated fat*, such as corn oil, has mostly polyunsaturated fatty acids. Polyunsaturated fats are liquid at room temperature; they stay liquid when chilled.

A fishy tale

The diet traditionally eaten in Mediterranean countries contains more fish than the diet most people eat in the UK. Mediterranean countries also have a lower incidence of coronary heart disease (CHD) than the Brits. Although other diet and lifestyle differences exist between the two, this correlation between high fish intake and low incidence of CHD sparked massive interest from the research world.

Some of the fats in oily fish such as salmon and sardines (called long chain omega-3 fatty acids) make tiny particles in blood called platelets less sticky, thus reducing the possibility that they'll clump together to form blood clots and plaques that might obstruct a blood vessel and trigger a heart attack. Omega 3s also help knock down levels of bad cholesterol and reduce high blood pressure, and help to keep the rhythm of the heart beat regular and keep arteries smooth and supple.

Most prestigious heart health organisations recommend the general population should be eating oily fish at least once a week. For people who've already had a heart attack the message is even firmer – following some major research studies in Wales and Italy, health professionals recommend that heart attack victims increase their intake of oily fish to two to three portions in the three months after their heart attack. This has been shown to reduce their risk of having another heart attack by almost one third. If they can't manage to eat that much fish then they should take a concentrated daily fish body oil supplement to provide approximately 1 gram of omega 3 per day. Besides, fish is also a good source of *taurine*, an amino acid that the journal *Circulation* notes helps maintain the elasticity of blood vessels, which means that the vessels may dilate to permit blood flow through.

The primary omega 3 is *alpha*-linolenic acid, which your body converts to hormone-like substances called *eicosanoids*. The eicosanoids eicosapentaenoic acid (EPA) and docosahexaenoic acid (DHA) reduce inflammation, perhaps by inhibiting an enzyme called COX-2, which is linked to inflammatory diseases such as rheumatoid arthritis. Many nutritionists and dietitians recommend that people suffering from rheumatoid arthritis increase their intake of omega-3 fatty acids.

As if omega 3s couldn't get any better, did we mention that they're also bone builders? Fish oils enable your body to create *calciferol*, a naturally occurring form of vitamin D, the nutrient that enables your body to absorb bone-building calcium – which may be why omega 3s appear to help hold minerals in bone – and increase the formation of new bone. And for our last word in praise of omega 3s, new research also suggests these super fats may even be involved in preventing dementia and depression; this may be due to large amount of omega 3 found in the structure of the brain.

If you simply take one message from all this information, take this one: eat more fish! If you really don't like fish, try some of the other sources of omega 3s that in the following list, or take fish body oil capsules as an alternative. Aim for about 0.5–1 gram of omega 3 per day. If you haven't eaten fish since your school dinner days, revisit it; things have certainly changed as far as delicious fish recipes go! If you're vegetarian, take a supplement – plant sources of omega 3 aren't used in the body in the same way as fish sources, so we can't be certain they offer the same potential benefits.

You can find omega 3s in:

- Broccoli
- Flaxseed or linseed oil
- Haddock
- Herring and kippers
- Kale
- Mackerel
- Pilchards
- Rapeseed oil
- Salmon
- Sardines
- Scallops
- Soya oil
- Spinach
- Trout
- Tuna (fresh, not tinned)
- Walnut oil

Before you shout, 'Waiter! I'll have the fish,' here's the other side of the coin. Native Alaskans, who eat plenty of fish, have a higher than normal incidence of haemorrhagic, or bleeding, strokes (a stroke caused by bleeding in the brain). A Harvard study found no significant link between fish consumption and bleeding strokes, but researchers say that more studies are needed.

Another potential downside reported in the newspapers is that fish (particularly oily fish harvested from inshore waters) may be contaminated with heavy metals like mercury and chemicals such as dioxins. However, in June 2004, the Food Standards Agency said that adults could safely enjoy up to four portions of oily fish per week (or two portions for pregnant or breast-feeding women). Pregnant or nursing mothers should avoid any intake of marlin, swordfish and shark, though, and have no more than two portions or around 140 grams of fresh tuna per week. This recommendation may be the smoke behind that media fire!

For the time being the benefits of fish seem to far outweigh any risks.

Delving deeper into fatty acids

If margarine is made from unsaturated fats such as sunflower and soybean oil, why is it solid? Because it's artificially saturated by food chemists who add hydrogen atoms to its unsaturated fatty acids. This process, known as *hydrogenation*, turns an oil, such as sunflower oil, into a solid fat that you can use in hard margarines, cakes, biscuits and in fried foods and takeaways.

A fatty acid with extra hydrogen atoms is called a *hydrogenated fatty acid*. Some hydrogenated fatty acids are also trans fatty acids. *Trans fatty acids* aren't healthy for your heart. Because of those added hydrogen atoms, they're more saturated, and they act like saturated fats, clogging arteries and raising the levels of cholesterol in your blood. To make it easier for you to control your trans fat intake, many food manufacturers have reduced the amount of trans in their products and often include how many grams of trans fats a product contains. The bottom line is: keep them to a minimum. The best way to do that is avoid too much fast food and takeaways, and cook from scratch, using unprocessed ingredients when you can.

The good news is that you can buy trans-fat-free margarines and spreads, including some that are made with plant sterols and stanols.

For those of you who want the whole story, plant *sterols* are natural compounds found in minute amounts in plant foods such as grains, fruits and vegetables, and soya beans. *Stanols* are compounds created by adding hydrogen atoms to sterols from wood pulp and other plant sources. Sterols and stanols can be concentrated and added to foods to enrich them. They work like little sponges, mopping up cholesterol in your intestines before it can make its way into your bloodstream. As a result, your total cholesterol levels and your levels of low-density lipoproteins (otherwise known as LDLs or *bad* cholesterol) go down. In some studies, two to three servings of stanol- or sterol-enriched foods a day can lower levels of bad cholesterol by between 10 and 15 per cent, with results showing up in as little as two weeks. Wow!

Some spreads, yoghurts, milk and drinks have added stanols and claim to help lower cholesterol when part of a healthy, balanced diet. Eating a healthy, balanced diet is the crucial bit – eating a whole load of processed food with the odd stanol-enriched food doesn't make a healthy diet!

Table 5-2 shows the kinds of fatty acids found in some common dietary fats and oils. Fats are characterised according to their predominant fatty acids. For example, the table shows that nearly 25 per cent of the fatty acids in corn oil are monounsaturated fatty acids. Nevertheless, because corn oil has more polyunsaturated fatty acid, corn oil is considered a polyunsaturated fatty acid. (Note for mathematicians: true, some of the totals in Table 5-2 don't add up to 100 per cent, because these fats and oils also contain other kinds of fatty acids in amounts so small that they don't affect the basic character of the fat.)

Table 5-2	What Fatty Acids Are in That Fat or Oil?			
Fat or Oil...	Saturated Fatty Acid (Excluding Trans) (%)	Monounsaturated Fatty Acid (%)	Polyunsaturated Fatty Acid (%)	Main Type of Fat or Oil
Sunflower margarine	16	20	41	Polyunsaturated
Rapeseed oil	7	60	30	Monounsaturated
Corn oil	13	24	59	Polyunsaturated
Olive oil	14	70	11	Monounsaturated
Palm oil	45	42	8	Saturated

Fat or Oil...	Saturated Fatty Acid (Excluding Trans) (%)	Monounsaturated Fatty Acid (%)	Polyunsaturated Fatty Acid (%)	Main Type of Fat or Oil
Safflower oil	10	13	72	Polyunsaturated
Soybean oil	14	23	57	Polyunsaturated
Butter	54	20	3	Saturated
Lard	39	45	11	Saturated*

*Because more than one third of its fats are saturated, nutritionists label lard a saturated fat.
Source: Royal Society of Chemistry & Ministry of Agriculture, Fisheries and Foods, McCance and Widdowson's, The Composition of Foods, 6th Ed, Royal Society of Chemistry, UK 1994

A diet high in saturated fats increases the amount of cholesterol circulating in your blood, which is believed to raise your risk of heart disease and stroke. A diet high in unsaturated fats reduces the amount of cholesterol circulating in your blood, which is believed to lower your risk of heart disease and stroke.

But here's a puzzle: Mother Nature's no fool. If cholesterol is uniformly bad, why does your body make so much of it? Read on.

Cholesterol: The Misunderstood Nutrient

Every healthy body *needs* cholesterol. Honest. Cholesterol is in and around your cells, in your fatty tissue, in your organs and in your glands. What's it doing there? Plenty of useful things. For example, cholesterol:

- Protects the integrity of cell membranes

- Helps enable nerve cells to send messages backwards and forwards

- Is a building block for vitamin D (a sterol), made when sunlight hits the fat just under your skin (for more about vitamin D, see Chapter 9)

- Enables your gall bladder to make *bile acids*, digestive chemicals that, in turn, enable you to absorb fats and fat-soluble nutrients such as vitamins A, D, E and K

- Is a base on which you build steroid hormones such as oestrogen and testosterone

Exploring the chemical structure of fatty acids

If you don't have a clue about the chemical structure of fatty acids, walking slowly with us through this explanation may be worth your while. The concepts are simple, and the information you find here applies to all kinds of molecules, not just fatty acids.

Molecules are groups of atoms hooked together by chemical bonds. Different atoms form different numbers of bonds with other atoms. For example, a hydrogen atom can form one bond with one other atom; an oxygen atom can form two bonds with other atoms; and a carbon atom can form four bonds to other atoms.

To actually see how this works, visualise a carbon atom as a round ball with holes in it. Your carbon atom (C) has four holes: one on top, one on the bottom and one on each side. If you stick a peg into each hole and attach a small piece of wood representing a hydrogen atom (H) to the peg on top, the peg on the bottom and the peg on the left, you have a structure that looks like this:

Methyl group

This unit, called a *methyl group*, is the first piece in any fatty acid. To build the rest of the fatty acid, you add carbon atoms and hydrogen atoms to form a chain. At the end, you tack on a group with one carbon atom, two oxygen atoms and a hydrogen atom. This group is called an *acid group*, the part that makes the chain of carbon and hydrogen atoms a fatty acid.

Saturated Fatty Acid

The preceding molecule is a *saturated fatty acid* because it has a hydrogen atom at every available carbon link in the chain. A *monounsaturated fatty acid* drops two hydrogen atoms and forms one double bond (two lines instead of one) between two carbon atoms. A *polyunsaturated fatty acid* drops more hydrogen atoms and forms several (poly) double bonds between several carbon atoms. Every hydrogen atom still forms one bond; every carbon atom still forms four bonds, but does it in a slightly different way. These sketches aren't pictures of real fatty acids, which have many more carbons in the chain and have their double bonds in different places, but they can give you an idea of what fatty acids look like up close. Not a pretty picture!

Instead of this:

You get this:

(a saturated fatty acid)　(a monosaturated fatty acid)

OR

(a polyunsaturated fatty acid)

Cholesterol and heart disease

Cholesterol has its uses but it has a dark side. Cholesterol makes its way into blood vessels, sticks to the walls and forms deposits that eventually block the flow of blood causing arteriosclerosis (clogging and hardening of the arteries), or can break away and cause a blood clot to form in an artery (thrombosis). The more cholesterol you have floating in your blood, the more cholesterol is likely to cross into your arteries, where it may increase your risk of heart attack or stroke.

Your GP can check your cholesterol levels. Doctors measure your cholesterol level by taking a sample of blood and measuring the amount of cholesterol in the small amount of blood they analyse. They can then work out how much cholesterol you'd have in 1 litre of blood. This is shown in millimoles per litre (mmol/l). When you get your result from the doctor, your total cholesterol level looks something like this: 4.8 mmol/l (that's a healthy level by the way!). Translation: you have 4.8 millimoles of cholesterol in every litre of blood.

As a general rule, an adult cholesterol level higher than 6.5 mmol/l is said to be a high risk factor for heart disease. A cholesterol level between 5.2 mmol/l and 6.5 mmol/l is considered a moderate risk factor. A cholesterol level below 5.2 mmol/l is considered a low risk factor. These figures only apply to adults because cholesterol targets for children don't exist in the UK.

Cholesterol readings can be affected by illness and can fluctuate from one day to another. This fluctuation is why doctors test for too much fat in the blood (hyperlipidaemia) three times before they make any diagnosis.

Finally, a word to the wise: cholesterol levels alone aren't the entire story. Many people with high cholesterol levels live to a ripe old age, but others with low total cholesterol levels develop heart disease. The reason is that total cholesterol is only one of several risk factors for coronary heart disease. Here are some more:

- A family history of heart disease
- Age (being older is riskier)
- An unfavorable ratio of lipoproteins (see the following section)
- Diabetes or high blood pressure
- High levels of homocystine (a substance produced by methionine in the body to repair damaged tissues)
- Lack of physical activity
- Sex (being male is riskier)
- Smoking
- Excess weight (especially if you carry your weight on your belly)

Living with lipoproteins

A *lipoprotein* is a fat (lipo = fat, remember) and protein particle that carries cholesterol through your blood. Your body makes four types of lipoproteins: chylomicrons, very low-density lipoproteins (VLDLs), low-density lipoproteins (LDLs) and high-density lipoproteins (HDLs). As a general rule, LDLs take cholesterol into blood vessels; HDLs carry it out of the body.

Lipoproteins start out big and fluffy. Then, as they keep moving, travelling around your body, they lose fat and end up small and dense. A lipoprotein is born as a *chylomicron*, made in your intestinal cells from protein and triglycerides. Chylomicrons are very low density, a term that means they have very little protein and lots of fat and cholesterol (protein is denser; that is, heavier and more compact than fat). After 12 hours of travelling through your blood and around your body, a chylomicron has lost virtually all of its fats. By the time the chylomicron makes its way to your liver, the only thing left is protein.

The liver, a veritable fat and cholesterol factory, collects fatty acid fragments from your blood and uses them to make cholesterol and new fatty acids. Your liver also keeps a check on the amount of cholesterol you get from food. If you eat more cholesterol, your liver may make less. If you eat less cholesterol, your liver may make more.

Okay, after your liver has made cholesterol and fatty acids, it packs them with protein as very low-density lipoproteins, which have more protein and are denser than chylomicrons. As VLDLs travel through your blood stream, they lose triglycerides, pick up cholesterol and turn into low-density lipoproteins. LDLs supply cholesterol to your body cells, which use it to make new cell membranes and manufacture sterol compounds such as hormones. That's the good news.

The bad news is that both VLDLs and LDLs are soft and squishy enough to pass through blood vessel walls. They carry cholesterol into blood vessels where it can cling to the inside wall, forming deposits, or *plaques*. These plaques may eventually block an artery, prevent blood from flowing through, and trigger a heart attack or stroke. That's the really bad news.

VLDLs and LDLs are sometimes called *bad cholesterol*, but this characterisation is a misnomer. They aren't cholesterol; they're just the rafts on which cholesterol sails into your arteries. Travelling through the body, LDLs continue to lose cholesterol. In the end, they become high-density lipoproteins, the particles sometimes called *good cholesterol*. Once again, this label is inaccurate. HDLs aren't cholesterol; they're simply protein and fat particles too dense and compact to pass through blood vessel walls, so they carry cholesterol out of the body rather than into arteries.

That's why a high level of HDLs may reduce your risk of heart attack regardless of your total cholesterol levels. Conversely, a high level of LDLs may raise your risk of heart attack, even if your overall cholesterol level is low. Hmmm, on second thoughts, maybe that does qualify them as *good cholesterol* and *bad cholesterol.*

Diet and cholesterol

Most of the cholesterol you need is made right in your own liver, which churns out about 1 gram (1,000 milligrams) a day from the raw materials in the proteins, fats and carbohydrates that you consume. But you also get cholesterol from food of animal origin: meat, poultry, fish, eggs and dairy products. Although some plant foods, such as coconuts and cocoa beans, are high in saturated fats, no plants actually have cholesterol. Dietary cholesterol has a relatively small influence on blood cholesterol, so rather than bore you with long lists of the cholesterol content of masses of foods we've compiled Table 5-3, which shows groups of foods classified as being high, moderate, low or cholesterol free. Eating a mixture of foods easily ensures that you get the 300 milligrams of cholesterol each day you need from dietary sources.

Plants don't have cholesterol, so no plant foods are on this list. No grains. No fruits. No veggies. No nuts and seeds.

Table 5-3	Sources of Dietary Cholesterol
Cholesterol Content	*Foods*
High	Liver, offal (and foods containing these such as pâté). Egg yolk, mayonnaise. Fish roes. Shellfish.
Moderate	Fat on meat, duck, goose and cold cuts such as salami. Full-fat milk, tinned milks, cream, ice cream, cheese, butter. Most manufactured pies, cakes, biscuits and pastries. Meat and fish products.
Low	Uncoated white and oily fish products, tinned fish in vegetable oil. Very lean meats, poultry (no skin). Skimmed milk, low-fat yoghurt, cottage cheese. Bread. Low-cholesterol margarine and spreads.
Cholesterol free	All vegetables and vegetable oils. Fruit including avocado and olives. Nuts. Cereals, pasta (without added egg), rice, popcorn (unbuttered). Egg white, meringue. Sugar.

A specialised cholesterol-lowering diet also exists: it's called the Portfolio Cholesterol-lowering Diet.

Research into how food might help lower cholesterol was prompted by looking at cholesterol levels in people who normally eat a lot of plant foods – these people tended to have much lower cholesterol levels than meat eaters.

Several foods came out as being potential stars of the cholesterol-lowering world, and when eaten together in combination, some scientists think they can be as powerful as drugs called statins to lower bad cholesterol (statin drugs are available over the counter, but doctors also prescribe them to a lot of patients with high cholesterol – in fact, since 1997 UK prescriptions for statins have increased 17 fold!).

The potential effect of eating the Portfolio Cholesterol-lowering foods seems very impressive: up to 20 per cent reduction in LDL bad cholesterol. But the diet is quite a tall order to follow and needs a big commitment. Table 5-4 outlines the diet.

Table 5-4	Main Portfolio Cholesterol-lowering Foods			
Food	*Active Cholesterol-lowering Ingredient*	*Recommended Daily Intake*	*Potential Cholesterol-lowering Effect*	*Foods Available*
Soya	Soya protein	25g or 3–4 servings	5–6% Total cholesterol reduction	Soya milk, yoghurt and custards, tofu, soya beans, soya nuts and flour
Soluble fibre (predominantly from oats)	Beta glucans	3g beta glucans from 3–4 servings of oats	5% reduction in total cholesterol	Oat cereals, breads and crackers
Plant sterols (special produce functional foods)	Sterol and stanols	2g plant sterols from combined products or from one yoghurt drink	Up to 10% reduction in bad cholesterol – similar to effect of statin drugs	Spreads, milks, yoghurts, oils and yoghurt drinks

Food	Active Cholesterol-lowering Ingredient	Recommended Daily Intake	Potential Cholesterol-lowering Effect	Foods Available
Nuts/almonds	Phenols, unsaturated fats, fibre and compounds like quercitin	30g unsalted nuts	5–10% total cholesterol reduction	

It is advisable to consult with your doctor or a dietitian before making any major change to your diet.

Chapter 6

Calories: The Energisers

- -

In This Chapter

▶ Discovering exactly what a calorie is

▶ Understanding where calories come from

▶ Explaining why men generally need more calories than women

▶ Finding out what happens when you get too many (or too few) calories

- -

Cars burn petrol to get the energy they need to move. Your body burns (*metabolises*) food to produce energy in the form of heat. This heat warms your body and powers every move you make.

Nutritionists measure the amount of heat produced by metabolising food in units called kilocalories (kcals). A *kilocalorie* is the amount of energy it takes to raise the temperature of one kilogram of water one degree centigrade at sea level. Energy is also expressed as kilojoules (kJ), the standard international (SI) unit for energy, and the more scientifically accurate way to express energy. However, most people are more familiar with food energy expressed as kilocalories. One kilocalorie is equal to 4.18 kilojoules.

In common use, nutritionists often substitute the word *calorie* for *kilocalorie*. This information isn't scientifically accurate: strictly speaking, a calorie is really $\frac{1}{1000}$ of a kilocalorie. But the word calorie is easier to say and easier to remember, so that's the term you sometimes see when you read about the energy in food and that's the word we use in this book. And few nutrition-related words have caused as much confusion and concern as the humble calorie. Read on to find out what calories mean to you and to your nutrition.

Counting the Calories in Food

When you read that a portion of food – say, one banana – has 105 calories, it means that metabolising the banana produces 105 calories of heat that your body can use for work.

You may wonder which kinds of food have the most calories. The answer is that it depends which nutrients they contain.

One gram of	Has this many calories
Fibre	2
Carbohydrate	3.75
Protein	4
Alcohol	7
Fat	9

In other words, a gram of protein or carbohydrate gives you fewer than half as many calories as a gram of fat. That's why high-fat foods, such as crisps, are higher in calories than low-fat foods, such as boiled potatoes.

Measuring the number of calories

Nutritionists measure the number of calories in food by actually burning the food in a *bomb calorimeter*, which is a box with two chambers – one inside the other. They weigh a sample of the food, put the sample on a dish and put the dish into the inner chamber of the calorimeter. They fill the chamber with oxygen and then seal it so that the oxygen can't escape. The outer chamber is filled with a measured amount of cold water, and the oxygen in the first chamber (inside the chamber with the water) is ignited by a heating element. When the food burns, an observer records the rise in the temperature of the water in the outer chamber. If the temperature of the water goes up 1 degree per kilogram of food, the food has 1 calorie; 2 degrees, 2 calories; or 250 degrees, 250 calories – or one small bar of chocolate!

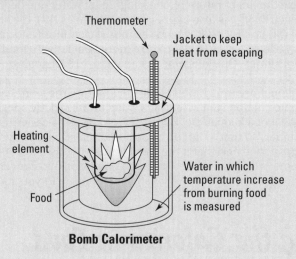

Thermometer

Jacket to keep heat from escaping

Heating element

Water in which temperature increase from burning food is measured

Food

Bomb Calorimeter

However, it's not always that simple, because foods usually contain a mix of nutrients. As a good example, take a chicken breast and a beefburger, both high-protein foods. If you serve the chicken without its skin, it contains very little fat, but the beefburger is (we hate to say) full of it. A 100-gram (3.5-ounce) skinless chicken breast provides 140 calories, and a 100 gram (3.5 ounce) burger has around 270 calories. (These are weights when cooked.)

Empty calories

All foods provide calories. All calories provide energy. But not all calories come with extra nutrients such as amino acids, fatty acids, fibre, vitamins and minerals. Some foods are said to give you *empty calories*. The term describes a calorie with few or no extra nutritional benefits.

The best-known empty-calorie foods are table sugar and *ethanol* (the alcohol found in beer, wine and spirits). On their own, sugar and ethanol give you energy – but no other nutrients. (See Chapter 7 for more about sugar and Chapter 8 for more about alcohol.)

People who drink too much alcohol can show nutritional deficiencies. Many alcoholic drinks deplete the body of B vitamins during their metabolism, but unlike most well-designed foods they don't compensate by bringing their own supply with them. The most common deficiency is of thiamin (vitamin B1), resulting in loss of appetite, depression, an inability to concentrate, nerve damage and, in its extreme form, a type of dementia. (For more on vitamin deficiency problems, check out Chapter 9.)

Of course, it's only fair to point out that sugar and alcohol are often ingredients in foods that do provide other nutrients. Sugar is found in fruit, and alcohol is found in wine – two very different foods that also both have a range of vitamins and minerals including antioxidant phytochemicals (see Chapter 11).

Even today in Western society, some people are malnourished because they can't afford enough food to get the nutrients they need. In the UK welfare food schemes and free school meals were introduced in 1944 to prevent malnutrition among poor schoolchildren. But many people who can afford enough food are nevertheless malnourished because they simply don't know how to choose a diet that gives them nutrients as well as calories. For these people, eating too many foods with empty calories can cause significant health problems such as heart disease, anaemia, weak bones, bleeding gums and even mental health problems including depression.

Every calorie counts

People who say that 'calories don't count' or that 'some calories count less than others' are usually trying to convince you to follow a diet that concentrates on one kind of food to the exclusion of most others. One common example that seems to arise like a phoenix in every generation of dieters is the *low-carbohydrate diet*.

The low-carbohydrate diet says to cut back or even entirely eliminate carbohydrate foods but eat as much fat and protein-rich food as you like. The idea is that this gives you some sort of metabolic advantage where the calories from these foods don't 'count'. Wouldn't it be wonderful if that were true? The problem is, it isn't. The facts are that all calories, regardless of where they come from, give you energy. If you take in more energy (calories) than you burn up each day, you gain weight. If you take in less than you use up, you lose weight. This nutrition rule is an equal-opportunity, one-size-fits-all fact that applies to everyone.

How Many Calories Do You Need?

Think of your energy balance as a bank account. You make deposits when you consume calories from food and drink. You make withdrawals when your body burns up energy on activity. Nutritionists divide the amount of energy you withdraw each day into two parts:

- ✔ The energy you need when your body is at rest
- ✔ The energy you need when you're physically active

To keep your energy account in balance, you need to take in enough each day to cover your withdrawals. As a general rule, infants and adolescents need more energy weight for weight than adults do, because they're continually making large amounts of new tissue. Similarly, an average man burns more energy than an average woman because his body is larger and has more muscle, thus leading to the totally unfair, but totally true fact that a 70-kilogram (11-stone) man can consume about 10 per cent more calories than a 70-kilogram (11-stone) woman of the same age and activity level, and still not gain weight. For the numbers, turn to the next section and Table 6-1.

Resting energy expenditures

Even when you're at rest, your body is busy burning calories. Your heart beats. Your lungs expand and contract. Your intestines digest food. Your liver processes nutrients. Your glands secrete hormones. Your muscles work gently. Cells send electrical impulses backwards and forwards among themselves, and your brain continually signals to every part of your body.

The energy that your resting body uses to do all this stuff is called (quite appropriately) *resting energy expenditure*, abbreviated as REE. The REE, also known as the *basal metabolic rate* or *BMR*, accounts for a whopping 60 to 70 per cent of all the calories you need each day. To find your resting energy expenditure (REE), you must first know your weight in kilograms (kg). One kilogram equals 2.2 pounds (lbs). So to get your weight in kilograms, first calculate your weight in pounds (1 stone is 14 pounds) then divide the amount in pounds by 2.2. For example, if you weigh 10 stone 10 pounds or 150 pounds, that's equal to 68.2 kilograms (150 ÷ 2.2). Enter that into the appropriate equation in Table 6-1 – and hey presto! You have your REE.

Table 6-1	How Many Calories Do You Need When You're Resting?
Sex and Age	**Use This Equation to Work Out Your REE**
Males	
0–3 years*	(60.9 × weight in kg) – 54
3–10 years*	(22.7 × weight in kg) + 495
10–17 years	(17.7 × weight in kg) + 657
18–29 years	(15.1 × weight in kg) + 692
30–59 years	(11.5 × weight in kg) + 873
60–74 years	(11.9 × weight in kg) + 700
75+	(8.4 × weight in kg) + 821
Females	
0–3 years*	(61.0 × weight in kg) – 51
3–10 years*	(22.5 × weight in kg) + 499
10–17 years	(13.4 × weight in kg) + 692
18–29 years	(14.8 × weight in kg) + 487
30–59 years	(8. 3 × weight in kg) + 846
60–74 years	(9.2 × weight in kg) + 687
75+	(9.8 × weight in kg) + 624

Source: Modified Schofield Equations from Dietary Reference Values for Food Energy and Nutrients for the United Kingdom. Report of the Panel on Dietary Reference Values of the COMA of Food Policy *(Department of Health, 1991)* * The National Research Council, Recommended Dietary Allowances *(Washington, D.C.: National Academy Press, 1989)*

Sex, glands and chocolate cake

A *gland* is an organ that secretes *hormones*, which are chemical substances that can change the function – and sometimes the structure – of other body parts. For example, your pancreas secretes *insulin*, a hormone that enables you to metabolise and store carbohydrates. At puberty, your sex glands secrete either the female hormones oestrogen and progesterone or the male hormone testosterone; these hormones trigger the development of secondary sex characteristics such as body and facial hair that make you look like either a man or a woman.

Hormones can also affect your REE. Your pituitary gland, a small structure in the centre of your brain, stimulates your thyroid gland (which sits at the front of your throat) to secrete hormones that influence the rate at which your tissues burn nutrients to produce energy.

When your thyroid gland doesn't secrete enough hormones (a condition known as *hypothyroidism*), you burn food more slowly and your REE drops. When your thyroid secretes excess amounts of hormones (a condition known as *hyperthyroidism*), you burn food faster and your REE is higher.

When you're frightened or excited, your adrenal glands (two small glands, one on top of each kidney) release *adrenaline*, the hormone that triggers your body's fight or flight responses. Your heartbeat increases. You breathe faster. Your muscles clench. And you burn food more quickly, converting it as fast as possible to the energy you need for the reaction. But these effects are temporary. The effects of the sex hormones, on the other hand, last as long as you live.

How your hormones affect your energy needs

If you're a woman, you know that your appetite can rise and fall in tune with your menstrual cycle. In fact, this fluctuation parallels what's happening to your REE, which goes up just before or at the time of ovulation. Your appetite is highest when menstruation starts, and then falls sharply. Being a man (and making lots of testosterone) makes satisfying your nutritional needs on a normal diet easier. Your male bones are naturally denser, so you're less dependent on dietary or supplemental calcium to prevent *osteoporosis* (severe loss of bone tissue) late in life. You don't lose blood through menstruation, so you need only two-thirds as much iron. That teenage boys develop wide shoulders and biceps while teenage girls get hips is no accident. Testosterone, the male hormone, promotes the growth of muscle and bone. Oestrogen promotes fat storage. As a result, the average male body has proportionally more muscle; the average female body, proportionally more fat.

Muscle is active tissue. It expands and contracts. It works. And when a muscle works, it uses more energy than fat (which insulates the body and provides a source of stored energy) but doesn't move an inch on its own. What this muscle versus fat battle means is that the average man's REE is about 10 per cent higher than the average woman's. In practical terms, that means that a 70-kilogram (11-stone) man can hold his weight steady while eating about 10 per cent more than a 70-kilogram (11-stone) woman who's of the same age and activity level.

No amount of dieting changes this unfair situation. A woman who exercises strenuously may reduce her body fat so dramatically that she no longer menstruates – an occupational hazard for endurance athletes. But she still has proportionately more body fat than an adult man of the same weight. If she eats what he does, and they perform the same amount of physical work, she still requires fewer calories than he to hold her weight steady.

Muscle weighs more than fat. This interesting fact is one that some people who take up exercise to lose weight discover by accident. One month into the weights and treadmill, their clothes fit better, but the scales register slightly higher because they've exchanged fat for muscle.

Energy for work

Your second largest chunk of energy is the energy you withdraw to spend on physical activity. That's everything from brushing your teeth in the morning to mowing the lawn or working out in the gym.

Your total energy requirement (the number of calories you need each day) is your REE plus enough calories to cover the amount of work you do.

Table 6-2 defines the energy expenditure of various activities ranging from the least energetic (sleeping) to the most (playing football, climbing stairs).

Table 6-2	How Active Are You When You're Active?	
Activity Level	**Activity**	**Average Energy Expenditure (Kcals per Hour) for a Person of 60 kg**
Resting	Sleeping, lying down, sitting, standing at rest, reading, writing, listening to radio, watching TV, eating	60–80
Very light	Sewing, dusting, cooking, playing a musical instrument, driving, washing up, ironing, playing snooker, bowling, general office and laboratory work	90–140

(continued)

Table 6-2 *(continued)*

Activity Level	Activity	Average Energy Expenditure (Kcals per Hour) for a Person of 60 kg
Light	Strolling gently at 2mph, vacuuming, washing and dressing, electrical work, painting and decorating, cricket	150–200
Moderate	Walking at 3mph, mopping floor, gentle gardening, cleaning windows, table tennis, sailing, golf, garage work, carpentry, bricklaying	210–260
Heavy	Brisk walking at 4mph, heavy gardening, dancing, moderate swimming, volleyball, slow jogging, labouring, digging, road construction, chopping wood, gentle cycling	270–350
Very heavy	Walking with a load uphill or cross country, brisk jogging, cycling, football, energetic swimming, tennis, skiing	360–600

Adapted from Dietary Reference Values for Food Energy and Nutrients for the United Kingdom. Report of the Panel on Dietary Reference Values of the COMA of Food Policy *(Department of Health, 1991)*

So How Much Should You Weigh?

A number of charts and tables claim to lay out *standard* or *healthy weights* for adults, but sometimes the figures appear so low that it seems you can't really get there without constant dieting.

Everyone knows overweight people who live long and happy lives and slim ones who die sooner than they should. However, people who are overweight have a higher risk of developing some illnesses, such as type 2 diabetes. Each person is made so differently that an ideal weight doesn't really exist. However, you need to know the healthy weight range for your height. One good guide is the *Body Mass Index (BMI)*, a number that measures the relationship between your weight and your height and offers some predictive estimate of your risk of weight-related disease. Another way is looking at your waist circumference. We look at both methods in the next sections.

Size matters: The biggest Brits

Data analysts Experian recently compiled a league table of areas of Britain based on hospital admissions for type 2 diabetes. Type 2 diabetes is a key indicator of obesity linked to poor diet and lack of physical activity and is in turn a major risk factor for heart disease. The analysts found a strong geographical divide in the results. The majority of the fattest areas are in the north of England and Wales. The ten leanest are all in southern England. Detailed analysis of the data shows that people in the leanest areas are much more likely to have the facilities, money and education to allow them to eat well and take regular exercise.

These inequalities must be addressed if the country is to reduce obesity and its associated medical problems successfully. Healthy food needs to be made more easily accessible to those on low incomes and exercise needs to be an affordable, achievable and safe option.

The ten fattest areas:

Hull

Knowsley, Merseyside

Blackburn, Lancashire

South Tyneside

Easington, County Durham

Merthyr Tydfil, Wales

Blaenau, Wales

Stoke on Trent

Pendle, Lancashire

Middlesbrough

The ten leanest areas:

Kingston upon Thames

Kensington and Chelsea

Westminster

Richmond upon Thames

Wandsworth

Isles of Scilly

Hammersmith and Fulham

Elmbridge, Surrey

Camden

South Buckinghamshire

Calculating your Body Mass Index

To calculate your BMI, measure your height in metres (take your shoes off) and figure out your weight in kilograms (definitely take your shoes off!). Now multiply the figure for your height by itself (square it). Divide your weight (in kilograms) by your height (in metres) squared.

For example, if you're 1.6 metres (5'3") tall and weigh 70 kilograms (11 stone) the equations for your BMI look like this:

$$\text{BMI} = \frac{70}{1.6 \times 1.6} = \frac{70}{2.6} = 27$$

Check out Table 6-3 to see how your BMI relates to your weight.

Table 6-3	Classification of Weight Categories using BMI
Category	**BMI**
Underweight	18.5
Healthy weight	18.5–24.9
Overweight	25–29.9
Moderately obese	30–34.9
Severely obese	35–39.9
Morbidly obese	>40

Current nutritional research suggests that the healthiest BMI is about 21.0. A BMI higher than 28 doubles the risk of illness (especially diabetes and heart disease) and death.

You can see that we give a range of weights for each BMI category. If you have a small frame and proportionately more fat tissue than muscle tissue (muscle is heavier than fat), you're likely to be at the low end of the range. When you have a large frame and proportionately more muscle than fat, you're likely to be at the high end. As a general (but by no means invariable) rule, that means that women – who have smaller frames and less muscle – weigh less than men of the same height and age.

BMI isn't a perfect measure of weight because it doesn't measure body fat specifically. If BMI alone is used, very muscular people can be mistakenly classified as obese.

Measuring your waist circumference

Looking at your waist circumference in conjunction with BMI is another useful method for assessing any weight-related health risks such as diabetes and heart disease. Fat stored centrally around the abdomen (tummy) – apple shaped or *central obesity* – is linked more closely to health risks than fat stored around the hips (pear shaped). Table 6-4 shows recommended cut off points. If you're in the 'at risk' category, avoid further weight gain. If you're in the 'substantially increased risk' category, we strongly advise you to lose weight. Adults of south Asian origin tend to have higher health risks factors at lower BMIs and waist circumferences than Western populations, so some organisations have proposed a lower waist circumference cut off for adults of south Asian origin of 90 centimetres for men.

Table 6-4	**Waist Circumference Cut-Off Points**	
	At Risk	*Substantially Increased Risk*
Men	> 94cm	>102cm
Women	>80cm	>88cm
South Asian men	>90cm	>102cm

Source: World Health Organisation, 1998

Looking closer at the weight/height relationship

Body Mass Index tables are so plentiful that you may think they're totally reliable in predicting who's healthy and who's not. But in reality they aren't always.

The problem is that real people and their differences keep entering into the equation. For example, the value of the BMI in predicting your risk of illness or death appears to be tied to your age. If you're in your 30s, a lower BMI is clearly linked to better health. If you're in your 70s or older, the evidence is less convincing that how much you weigh plays a significant role in determining how healthy you are or how much longer you'll live. In between, from age 30 to age 74, the relationship between your BMI and your health is, well, in between – more important early on, less important later in life.

If the clothes fit

A recent study in Scotland suggested clothing size could be a useful way to help decide when being overweight starts to become a health risk. Researchers looked at clothing size, body size and health risk in over 350 adults. Unsurprisingly, they found a strong link between clothing size, BMI and waist circumference. However, using a mathematical model they were able to link a higher risk of heart disease with size 36 or above trousers for men and a dress size of 16 or above for women. Having trousers over waist size 38 in men indicated a nearly four-fold chance of heart disease. For women, a dress size 18 or above meant they were seven times more likely to have at least one of the main risk factors for heart disease. Ninety-eight per cent of people in the study knew their garment size, suggesting this can be a useful indicator that people can use to tell whether their health is at risk.

Many nutritionists focus not only on weight/height (the BMI), but on the importance of other factors including how fit you are and how good the protective qualities of your diet are.

Here are some interesting extra dimensions in the weight/health relationship:

✔ People who are overweight may be more prone to ill health because they do less physical activity and so are less fit. Very good evidence shows that improving fitness in overweight people is effective in improving their health. Overweight people who are active reduce their risk of disease and death more than overweight people who are inactive – and the more active they are, the lower the risk.

✔ People who are overweight may be more likely to be unwell because they eat an energy-dense but poor-quality diet (based on foods high in saturated fat, sugar and salt, and low in antioxidants and fibre). In this case the remedy may be to improve the protective quality of the diet by eating more fruit and vegetables, pulses, wholegrain cereals, nuts and fish.

Adding to the confusion is the fact that an obsessive attempt to lose weight may itself be hazardous to your health (see Chapter 14). Thirty-four million people try to lose weight in the UK every year and people spend £10 billion (yes, you read that right) on diet clubs, special foods and over-the-counter remedies aimed at weight loss. The UK has the second highest spend on dieting in Europe (after Germany) – working out at a massive £300 per slimmer per year. Very often the fad diets, the pills and the special foods don't work, which may leave dieters feeling worse and heavier than they did before they started. A recent Weightwise survey of 4,000 UK adults trying to lose weight showed that only 40 per cent were able to follow the diet for more than a month and 90 per cent regained the weight they lost. Shockingly, one in five 'slimmers' ended up more than a stone heavier than when they started.

Some drugs that claim to help with weight loss may be dangerous. For example, ephedra and some other so-called diet pills such as yerba and guarana are stimulants and have been linked to serious adverse side effects such as increased blood pressure, headache, gastrointestinal upset and psychiatric disturbances such as increased anxiety and nervousness. These drugs are banned in some countries.

Controlling your weight effectively

Remember that every body is built differently and each person has different calorie needs. The guidelines in this book aren't hard-and-fast rules.

However, realistic rules can enable you to control your weight safely and effectively:

- ✔ **Rule No. 1: Not everybody starts out with the same set of genes – or fits into the same pair of jeans.** Some people are naturally larger and heavier than others. If that's you, and your weight is within the healthy range, don't waste time trying to fit in with someone else's idea of perfection. Relax and enjoy your own body. Spot reduction on fatty areas is virtually impossible by dieting.

- ✔ **Rule No. 2: If you're overweight and your doctor agrees with your decision to diet, you don't have to set world records to improve your health.** Even a moderate loss of weight can be highly beneficial. Losing just 10 to 15 per cent of your body weight can lower high blood sugar, high cholesterol and high blood pressure, reducing your risks of diabetes, heart disease and stroke. For a 90-kilogram person that's just 9 kilograms!

- ✔ **Rule No. 3: The only number you need to remember is *3,500*, the number of calories it takes to gain or lose 0.5 kilograms (1 pound) of body fat.** So if you simply:

  - Cut your calorie consumption from 3,000 calories a day to 2,500 and continue to do the same amount of physical work, you'll lose 0.5 kilograms a week.

  - Go the other way, increasing from 2,500 to 3,000 calories a day, without increasing the amount of work you do, and seven days later you'll be 0.5 kilograms heavier.

Losing Weight by Controlling Calories

Modest calorie restrictions can produce safe and beneficial weight loss. You simply eat a sensible diet that includes a wide variety of different foods containing sufficient amounts of essential nutrients. The Department of Health has an interesting set of estimated average energy requirements for adult men and women based on actual measurements of the amount of daily calories burned by healthy adults of different ages and weights at different levels of activity. Table 6-5 shows the estimated average requirements for energy (kcals per day) for adult men and women based on age, weight and activity level. You can see that if you're active you can take in more calories and still keep your weight steady. If you're less active, you need fewer calories.

But here's a tip for smart calorie counters: not only does physical activity help to protect against heart disease and cancer, but when you're looking to lose a few pounds, increasing your activity is better than just cutting back on calories.

Table 6-5   Estimated Average Calorie Requirements for Energy (Kcals/Day) for Adults Based on Age, Weight and Activity Level

Weight (kg)	Inactive	Moderately Active	Weight (kg)	Inactive	Moderately Active
Men 19–29 years			**Women 18–29 years**		
60	2,224	2,727	50	1,722	1,961
65	2,344	2,846	55	1,818	2,081
70	2,440	2,990	60	1,937	2,200
75	2,559	3,110	65	2,033	2,320
80	2,655	3,229	70	2,129	2,440
Men 30–59 years			**Women 30–59 years**		
65	2,272	2,751	50	1,746	2,009
70	2,341	2,846	55	1,818	2,081
75	2,440	2,966	60	1,866	2,129
80	2,511	3,062	65	1,913	2,200
85	2,583	3,157	70	1,985	2,272
Men 60–64 years			**Women 60–64 years**		
74	2,380	n/a	63.5	1,900	n/a
Men 65–74 years			**Women 65–74 years**		
71	2,330	n/a	63	1,900	n/a
Men over 75 years			**Women over 75 years**		
69	2,100	n/a	60	1,810	n/a

Source: Adapted from Dietary Reference Values for Food Energy and Nutrients for the United Kingdom. Report of the Panel on Dietary Reference Values of the Committee on Medical Aspects of Food Policy (Department of Health, 1991). N/a = data not availableInvestigating Over-the-counter Weight Loss Drugs

Huge amounts of time and money have been put into finding drug treatments to help in weight loss. However, at the time of writing only one drug that's been sufficiently tried and tested for both safety and effectiveness is available in the UK. Xenical is available on prescription from a GP as Orlistat or over the counter as Alli.

The active ingredient in the drug attaches to the fat-digesting enzymes in your gut and prevents them from digesting about a quarter of the fat in your food. Because the gut can't absorb undigested fat, it passes out of your body instead of turning into calories.

You take the drug three times a day with meals containing no more than 15 grams of fat. Going above this level leads to unpleasant side effects including wind, stomach cramps and loose fatty or oily stools, sometimes with oily spotting and sudden bowel motions.

Xenical is licensed for use in adults over 18 with a BMI of 28 or more but only up to six months. If you take it you're advised to also take a multivitamin (containing fat-soluble vitamins A, D, E and K) once a day, at bedtime, to avoid possible vitamin deficiency.

The drug isn't a magic cure and only allows small amounts of additional weight loss over and above that achieved from a low-fat calorie reduced diet. It's not a replacement for a calorie-controlled diet and won't deal with any underlying psychological reasons for weight gain. Some people find it more useful than others, but it can help to encourage healthier eating patterns.

The trick is managing your calories and not letting them manage you. When you know that fats are more fattening than proteins and carbohydrates, and that your body burns food to make energy, you can strategise your energy intake to match your energy expenditure, and vice versa. Here's how: turn straight to Chapter 13 to find out about a healthy diet and Chapter 14 to find out about planning nutritious meals.

Chapter 7

Carbohydrates: A Complex Story

*C*arbohydrates (the name means carbon plus water) are sugar compounds made by plants when the plants are exposed to light. This process of making sugar compounds is called *photosynthesis*, from the Latin words for 'light' and 'putting together'.

In this chapter we shine the light on the different kinds of carbohydrates, illuminating all the nutritional nooks and crannies to explain how each contributes to your health, not to mention being a delicious daily staple.

Considering Carbohydrates

Carbohydrates come in three varieties: simple carbohydrates, complex carbohydrates and dietary fibre. All are composed of units of sugar. What makes one carbohydrate different from another is the number of sugar units it contains and how the units are linked together.

 ✔ **Simple carbohydrates** are carbohydrates with only one or two units of sugar.

 • A carbohydrate with one unit of sugar is called a *simple sugar* or a *monosaccharide* (*mono* = one; *saccharide* = sugar). Fructose (fruit sugar) is a monosaccharide, and so are glucose (blood sugar; the sugar produced when you digest carbohydrates) and galactose (one of the sugars derived from digesting lactose; milk sugar).

 • A carbohydrate with two units of sugar is called a *double sugar* or a *disaccharide* (*di* = two). Sucrose (table sugar), which is made of one unit of fructose and one unit of glucose, is a disaccharide.

✔ **Complex carbohydrates**, also known as *polysaccharides* (*poly* = many), have more than two units of sugar linked together. Carbs with three to eight units of sugar are sometimes called *oligosaccharides* (*oligo* = many).

- Raffinose is a *trisaccharide* (*tri* = three) that's found in potatoes, beans and beetroot. It has one unit each of galactose, glucose and fructose.

- Stachyose is a *tetrasaccharide* (*tetra* = four) found in the same vegetables we mention in the previous item. It has one fructose unit, one glucose unit and two galactose units.

- Starch, a complex carbohydrate in potatoes, pasta and rice, is a definite polysaccharide, made of many units of glucose.

✔ **Dietary fibre** is a term that distinguishes the fibre in food from the natural and synthetic fibres (silk, cotton, wool, nylon) used in fabrics. Dietary fibre is a third kind of carbohydrate.

- Like the complex carbohydrates, dietary fibre (cellulose, hemicellulose, pectin, beta-glucans, gum) is a polysaccharide. Lignin, a different kind of chemical, is also called a dietary fibre.

- Some kinds of dietary fibre also contain units of *uronic acids*, chemicals derived from fructose, glucose and galactose. (Galacturonic acid, from galactose, is an example of an uronic acid.)

Dietary fibre isn't like other carbohydrates. Human digestive enzymes can't break the bonds that hold its sugar units together. Although the bacteria living naturally in your intestines convert very small amounts of dietary fibre to fatty acids. r (For more about fatty acids, see Chapter 5.)

The way that fibre is classified varies around the world depending on the scientific method used to measure it in foods. A form of fibre receiving a lot of attention at the moment is *resistant starch* (starch that resists digestion). This type of fibre is only digested by friendly bugs in the colon and so may improve gut health. We'll probably see more of resistant starch in the future in the form of *functional foods* (foods specifically designed to offer a particular health benefit; see Chapter 18 for more).

In the next section we talk about how your body gets energy from carbohydrates. Because dietary fibre doesn't provide much energy, we'll put it aside for the moment and get back to it in the 'Dietary Fibre: The Non-nutrient in Carbohydrate Foods' section, later in this chapter.

Charting the sweetness of carbs

The information in the following table is trivia for your own personal nutrition data bank. It helps to illustrate how different types of carbs are formed. The simpler sugars like monosaccharides and disaccharides tend to be sweeter and are found naturally in foods; the polysaccharides are much more complex molecules. Some (pectin) are found naturally in foods and others (hemicellulose) are found in plant cell walls but are also manufactured for a variety of both food and non-food products. The bottom line is that the *saccharide* bit in the name tells you that the sugar is a carbohydrate, and the *mono*, *di* or *poly* bit tells you whether it's made up of one, two or many sugars.

Naming the Sugar Units in Carbohydrates

Carbohydrate	*Composition*
Monosaccharides (one sugar unit)	
Fructose (fruit sugar)	One unit of fructose
Galactose (made from lactose [milk sugar])	One unit of galactose
Glucose (sugar unit used for fuel)	One unit of glucose
Disaccharides (two sugar units linked together)	
Lactose (milk sugar)	Glucose + galactose
Maltose (malt sugar)	Glucose + glucose
Sucrose (table sugar)	Glucose + fructose
Polysaccharides (many sugar units linked together)	
Cellulose	Many glucose units
Gums	Mainly galacturonic acid
Hemicellulose	Arabinose* + galactose + mannose* + xylose** + uronic acids
Pectin	Galactose + arabinose + galacturonic acid
Raffinose	Galactose + glucose + fructose
Stachyose	Glucose + fructose + galactose + galactose
Starch	Many glucose units

* This sugar is found in many plants.
** This sugar is found in plants and wood.

Carbohydrates and energy: a biochemical love story

Your body runs on glucose, the molecules your cells burn for energy. (For more information on how you get energy from food, check out Chapter 6.)

Proteins, fats and alcohol also provide energy in the form of calories. And protein does eventually give you glucose, but it takes a long time, relatively speaking, for your body to get the glucose.

But all the digestible carbohydrates you get from food provide either glucose or sugar units that your body can convert quickly to glucose. The glucose is then carried into your cells with the help of *insulin*, a hormone your pancreas secretes.

How glucose becomes energy

Inside your cells, the glucose from carbohydrates is burned to produce heat and *adenosine triphosphate* (ATP), a molecule that stores and releases energy as required by the cell. The transformation of glucose into energy occurs with oxygen or without it.

- ✔ Glucose is converted to energy with oxygen in the *mitochondria*, tiny bodies in the jelly-like substance inside every cell. This conversion yields energy (ATP and heat) plus water and carbon dioxide, a waste product.

- ✔ Red blood cells don't have mitochondria, so they change glucose into energy without oxygen. This yields energy (ATP and heat) and lactic acid.

- ✔ Glucose is also converted to energy in muscle cells. Muscle cells have mitochondria, so they can process glucose with oxygen. But if the level of oxygen in the muscle cell falls very low, the cells can just go ahead and change glucose into energy without it. This is most likely to happen when you exercise so strenuously that you (and your muscles) are, literally, out of breath.

Being able to turn glucose into energy without oxygen is a handy trick, but here's the downside: one by-product is lactic acid. Why does that matter? Too much lactic acid makes your muscles ache.

Having too much of a good thing

Your cells use energy very carefully. Any glucose the cell doesn't need for its daily work is joined with other glucose cells and converted to *glycogen* (animal starch) and tucked away as stored energy in your liver and muscles. Glycogen is often called animal starch because it's the form of carbohydrate that only exists within animal cells; all other carbohydrates are produced within plant cells.

Your body can pack about 400 grams (14 ounces) of glycogen into liver and muscle cells. A gram of carbohydrates, including glucose, has 3.75 calories. If you add up all the glucose stored in glycogen to the small amount of glucose in your cells and blood, it equals about 1,800 calories of energy.

If your diet provides more carbohydrates (or any other energy source) than you need to produce this amount of stored calories in the form of glucose and glycogen in your cells, blood, muscles and liver, your body converts the excess into fat. If you really love your carbs, the way to combat expanding hips or any other area is to get more active and increase your need for energy – the bigger your need for energy, the bigger your plate!

Other ways your body uses carbohydrates

Providing energy is an important job, but it isn't the only thing carbohydrates do for you. Carbohydrates also protect your muscles. When you need energy, your body looks for glucose from carbohydrates first. If none is available because you're on a carbohydrate-restricted diet, for example, your body begins to rely on glucose stored as glycogen in fatty tissue and then moves on to burning its own protein tissue (muscles). If this use of proteins for energy continues long enough, you'll run out of fuel and die.

A diet that provides sufficient amounts of carbohydrates keeps your body from eating its own muscles. That's why a carbohydrate-rich diet is sometimes described as *protein sparing*.

Carbohydrates also:

   ✔ Provide nutrients for the friendly bacteria in your intestinal tract that help digest food.

   ✔ Assist in your body's absorption of calcium.

   ✔ May help lower cholesterol levels and regulate blood pressure. These effects are special benefits of dietary fibre, which we discuss in the 'Dietary Fibre: The Non-nutrient in Carbohydrate Foods' section, later in this chapter.

Finding the carbohydrates you need

The most important sources of carbohydrates are plant foods – fruits, vegetables and grains. Milk and milk products contain the carbohydrate lactose (milk sugar), but meat, fish and poultry have no carbohydrates at all.

The dietary reference values (DRVs) for food energy and nutrients don't set specific recommendations for the amount of carbohydrate foods you should eat. However, most nutritionists recommend that 50 per cent of your daily calories need to come from carbohydrate foods. The Eatwell plate (see more about that in Chapter 14) makes it easy for you to build a nutritious, carb-based diet with portion allowances based on how many calories you consume each day in 6 to 11 servings of complex carbohydrate foods (bread, cereals, pasta, rice and potatoes), plus at least five portions of fruit and vegetables.

In nutritional food group terms, potatoes are classified as complex carbohydrates even though they're a vegetable. This is because potatoes are a staple food in the UK diet, and although they do contain some of the nutrients (like vitamins and minerals) you look for in fruit and vegetables, their main nutritional role is to provide energy. So we're sorry but you can't count your spuds as one of your five a day!

The foods we have talked about so far provide simple carbohydrates, complex carbohydrates and the natural bonus of dietary fibre. Table sugar, honey, soft drinks and sweets provide simple carbohydrates but few other nutrients, and dietitians recommended you consume these carbs on an occasional basis only.

One gram of carbohydrates has 3.75 calories. To find the number of calories from the carbohydrates in a serving, multiply the number of grams of carbohydrates by 3.75. For example, one bread roll has about 30 grams of carbohydrates, equal to about 112 calories (30×3.75). Remember, that number doesn't account for all the calories in the serving. The bread roll may also contain at least some protein and fat, and these two nutrients add calories.

Some problems with carbohydrates

Some people have a hard time handling carbohydrates. For example, if you have type 1 diabetes, your pancreas doesn't produce enough insulin to carry all the glucose produced from carbohydrates into your body cells. If you have type 2 diabetes you can produce insulin but your cells are resistant to it. As a result, the glucose continues to circulate in your blood until it's excreted through the kidneys. One way to tell whether someone has diabetes is to test the level of sugar in that person's urine.

Other people can't digest carbohydrates because their bodies lack the specific enzymes needed to break the bonds that hold a carbohydrate's sugar units together. For example, many (some say most) adult Asians, Africans, Middle Easterners, South Americans and Eastern, Central or Southern Europeans are deficient in *lactase*, the enzyme that splits lactose (milk sugar) into glucose and galactose. If these people drink milk or eat milk products, they end up with a lot of undigested lactose in their intestinal tracts. This undigested lactose makes the bacteria living there happy as sandboys – but not the person who owns the intestines: because bacteria feast on the undigested sugar, they excrete waste products that give their host diarrhoea, wind and stomach pain.

To avoid this problem many national cuisines purposely avoid using fresh animal milk as an ingredient. (Quick! Name one native Asian dish that's made with milk. No, coconut milk doesn't count.) Does that mean that people living in these countries don't get enough calcium? No. They use low-lactose dairy foods such as yoghurt or paneers (a type of soft cheese) or simply substitute other high-calcium foods such as green vegetables and seeds. Soya products such as tofu and soya dairy alternatives are often calcium enriched.

No carbs, low carbs or slow carbs?

Carbs get quite a bashing these days from the diet industry and the media. For years all people ever talked about was cutting down on fat when it came to losing weight, but then an explosion of low-carb diets hit the bookshelves and the headlines. Low-carb diets sparked massive controversy and furious debate in the nutrition world. Some people following the diets lose weight with no side effects; others lose weight but feel awful with constipation, headaches and nausea. Nutritionists have justified concerns over replacing carbs with protein and fat with regard to heart health and increased risk of kidney disease. The jury is still out on how safe low-carb/high-protein/high-fat diets really are.

The low carbs controversy has generated significant research into how people digest and absorb carbs, and we now know that all carbs weren't created equal. Some complex carbs, such as wholemeal bread and jacket potatoes, that had always been recommended in weight-loss diets were shown to break down to glucose very quickly, causing a spike in blood glucose and a resultant surge in insulin release to mop up the excess glucose and store it as energy. After the insulin does its job, blood glucose falls again and you feel hungry again. Other carbohydrate foods such as oats, new potatoes and pulses break down to glucose much more slowly and keep blood glucose levels a lot more stable.

The glycaemic index (GI) is the classification we use to identify which carbs are quickly broken down to glucose (high GI) and which are slowly broken down (low GI). The research is still ongoing, but evidence suggests that understanding the different rates at which carbs are digested may be a useful tool in the fight against obesity, heart disease and diabetes. Instead of talking about no carbs or low carbs, the real answer may lie in slow carbs.

A second solution for people who don't make enough lactase is to use a *predigested milk product* such as yogurt, buttermilk or sour cream, all made by adding friendly bacteria that digest the milk (that is, break the lactose apart) without spoiling it. Other solutions include lactose-free cheeses and enzyme-treated milk.

Who needs extra carbohydrates?

The small amount of glucose in your blood and cells provides the energy you need for your body's daily activities. The 400 grams of glycogen stored in your liver and muscles provides enough energy for ordinary bursts of extra activity.

But what happens when you have to work harder or longer than that? What if you're a marathon runner who uses up your available supply of glucose before you finish your competition? (That's why marathoners often run out of fuel at 20 miles, 6 miles short of the finish line, a phenomenon called *hitting the wall*.)

After your body exhausts its supply of glucose, including the glucose stored in glycogen, it starts pulling energy first out of fat and then out of muscle. But extracting energy from body fat requires large amounts of oxygen, which is likely to be in short supply when your body has run 20 miles. So athletes have to find another way to leap the wall: they eat carefully chosen carbs and in bigger quantities.

The general guide for athletes is to achieve a carbohydrate intake of between 6 and 10 grams per kilogram of body weight, or around 70 per cent of energy intake. This should be predominantly made up from wholegrain, slow-release carbs (low GI; see the sidebar 'No carbs, low carbs or slow carbs') such as wholegrain bread, pasta, oats and barley. After intensive training athletes may need quick-releasing carbs such as ripe bananas, jam sandwiches and chocolate (high GI) to replace glycogen stores from the muscles and liver. This need is why many sports drinks are quite sugary, and you often see an athlete eating chocolate after an event. Consuming sugar during the race also gives you extra short-term bursts of energy because the body rapidly converts sugar to glycogen and carries it to the muscles.

Don't consume *straight sugar* (such as sweets or honey) during exercise because it's *hypertonic*. Using straight sugar can increase dehydration and make you nauseated. Get the sugar you want from sweetened *isotonic* or *hypotonic* drinks, which provide energy as well as fluid. The label on the sports drink also tells you whether the liquid contains salt (sodium chloride) to help the absorption of the fluid.

The enzyme name game

An interesting nutritional snippet is that the names of all enzymes end in the letters *ase*. An enzyme that digests a specific substance in food often has a name similar to the substance but with the letters *ase* at the end. For example, *proteases* are enzymes that digest protein; *lipases* are enzymes that digest fats; *lactase* is the enzyme that digests lactose. Simple, eh?

A quick summary to help you make head or tail of the range of sports drinks available:

- **Hypertonic drinks** usually have more than 10 grams of carbohydrate per litre and inhibit fluid absorption. They aren't generally recommended during exercise unless you're sufficiently hydrated and simply want to take in more energy.

- **Hypotonic drinks** usually contain less than 4 grams of carbohydrate per litre and help rehydration, especially if they have added sodium. However, they don't provide a significant contribution to energy needs.

- **Isotonic drinks** usually contain about 4–8 grams of carbohydrate per litre. This is enough to provide a small amount of carbohydrate as immediate energy during exercise but not so much that it prevents you from absorbing the fluid. These drinks can be useful before, during or after exercise.

Dietary Fibre: The Non-nutrient in Carbohydrate Foods

Dietary fibre is a group of complex carbohydrates that aren't a source of energy for human beings. Because human digestive enzymes can't break the bonds that hold fibre's sugar units together, fibre adds no calories to your diet and your body can't convert fibre to glucose.

Ruminants (animals, such as cows, that chew the cud and have several stomachs) have a combination of digestive enzymes and digestive microbes that enable them to extract the nutrients from insoluble dietary fibre (cellulose and some hemicelluloses). But not even these creatures can break down *lignin*, an insoluble fibre in plant stems and leaves and the predominant fibre in wood.

Fibre factoid

The amount of fibre in a serving of food may depend on whether the food is raw or cooked. For example, a 100 gram serving of plain dried prunes has 5.7 grams of fibre and a 100 gram serving of cooked tinned prunes in syrup has 2.8 grams of fibre.

When you stew and tin prunes, they absorb water and sugar and plump up. The syrup adds weight but no fibre. So a serving of prunes-plus-syrup has slightly less fibre per gram than a same-weight serving of plain dried (ready to eat) prunes.

Just because you can't digest dietary fibre doesn't mean it isn't a valuable part of your diet. The opposite is true. Dietary fibre is valuable *because* you can't digest it!

The two kinds of dietary fibre

Nutritionists classify dietary fibre as either *insoluble* fibre or *soluble* fibre, depending on whether it dissolves in water. (Both kinds of fibre resist human digestive enzymes.)

✔ **Insoluble fibre,** such as cellulose, some hemicelluloses and lignin, is found in whole grains and other plants. This kind of dietary fibre is a natural laxative. It absorbs water, helps you feel full after eating and stimulates your intestinal walls to contract and relax. These natural contractions, called *peristalsis*, move solid materials through your digestive tract.

By moving food quickly through your intestines, insoluble fibre may help prevent or relieve digestive disorders such as constipation or diverticulosis (an infection caused by food getting stuck in small pouches in the wall of your colon). Insoluble fibre also bulks up your faeces and makes them softer, reducing your risk of developing haemorrhoids and lessening the discomfort if you already have them.

✔ **Soluble fibre,** such as pectins in apples and beta-glucans in oats and barley, seems to lower the amount of cholesterol circulating in your blood (your *cholesterol level*). This may be why a diet rich in fibre appears to offer some protection against heart disease.

A benefit for dieters: soluble fibre forms gels in the presence of water, which is what happens when apples and oat bran reach your digestive tract. Like insoluble fibre, soluble fibre can make you feel full without adding calories.

Your body doesn't absorb ordinary soluble dietary fibre because it can't digest it. But in 2002 researchers in the USA fed laboratory mice a form of soluble dietary fibre called *modified citrus pectin*. The fibre, which comes from citrus fruit, can be digested. When fed to laboratory rats, it appeared to reduce the size of tumours caused by implanted human breast and colon cancer cells. The researchers believe that the fibre prevents cancer cells from linking together to form tumours. Plenty more research is needed before we can make firm conclusions, but this discovery does seem to add fuel to the message to eat more fruit and veg!

Getting fibre from food

You find fibre in all plant foods – fruits, vegetables and grains. But you find absolutely no fibre in foods from animals – meat, fish, poultry, milk, milk products and eggs.

A balanced diet with lots of foods from plants gives you both insoluble and soluble fibre. In general terms soluble fibre helps to reduce glucose and cholesterol absorption, and insoluble fibre speeds up the time it takes food to pass through the gastrointestinal tract. Most foods that contain fibre have both kinds, although the balance usually tilts towards one or the other. For example, the predominant fibre in an apple is pectin (a soluble fibre), but apple peel also has some cellulose, hemicellulose and lignin.

Table 7-1 shows you which foods are the best sources of what fibre. A diet rich in plant foods gives you adequate amounts of dietary fibre.

Table 7-1	Fibre Facts
Fibre	*Found In . . .*
Soluble fibre	
Beta-glucans	Oats, barley
Gums	Beans, cereals (oats, rice, barley), seeds, seaweed
Pectin	Fruits (apples, strawberries, citrus fruits)
Insoluble fibre	
Cellulose	Leaves (cabbage, spring greens), roots (carrots, parsnips), bran, whole wheat, rye, peas, beans
Hemicellulose	Seed coverings (bran, whole grains)
Lignin	Plant stems, leaves and skin

How much fibre do you need?

The recommended dietary reference value (DRV) for adults is 18 grams of fibre a day. Children's intake should be proportionately lower than that of adults. The amounts of dietary fibre recommended by DRVs are believed to give you the benefits you want without causing fibre-related unpleasantries.

If you eat more than enough fibre, your body tells you right away. All that roughage may irritate your intestinal tract, which will issue an unmistakable protest in the form of intestinal gas or diarrhoea. In extreme cases, if you don't drink enough fluid to carry the fibre you eat easily through your body, the dietary fibre may form a mass that can end up as an intestinal obstruction (for more about water, see Chapter 12).

If you decide to increase the amount of fibre in your diet, follow our advice:

✔ Do it *very* gradually, a little bit more every day. That way you're less likely to experience intestinal distress. In other words, if your current diet is heavy on no-fibre foods such as meat, fish, poultry, eggs, milk and cheese, and low-fibre foods such as white bread and white rice, don't load up on bran cereal (36.4 grams dietary fibre per 100 gram serving) or dried figs (6.9 grams per serving) all at once. Start by adding a serving of cornflakes (0.9 grams dietary fibre) at breakfast one day, then maybe an apple (1.8 grams) at lunch the following day, a pear (2.2 grams) at mid-afternoon later in the week and a small tin of baked beans (6.9 grams) at dinner. Four simple additions, and already you're up to 18 grams of dietary fibre.

✔ Follow the recommendations of the Eatwell plate (see Chapter 14) and increase your consumption of whole grain products, vegetables and fruits – all good sources of dietary fibre.

✔ Always check the nutrition label whenever you shop (for more about food labelling, see Chapter 14). When choosing between similar products, just take the one with the higher fibre content per serving. For example, white pitta bread generally has about 2.2 grams of dietary fibre per serving. Wholemeal wheat pitta bread has 5.8 grams. From a fibre standpoint, you know which works better for your body. Go for it!

By the way, dietary fibre is like a sponge. It soaks up liquid, so increasing your fibre intake may deprive your cells of the water they need to perform their daily work. Unless you're already drinking at least six large glasses of water every day, we and any other dietitians worth their salt suggest upping your fluid intake when you consume more fibre.

Table 7-2 shows the amounts of all types of dietary fibre (insoluble plus soluble) in a 100-gram serving of specific foods. By the way, nutritionists like to measure things in terms of 100-gram portions because that makes comparing foods at a glance possible. To find out how much fibre you're eating in a single serving you can simply look at the nutrition label on the side of the package that gives the nutrients per portion – much easier!

Remember that the amounts on Table 7-2 are averages. Different brand-name processed products (breads, cereals, cooked fruits and vegetables) may have more or less fibre per serving.

Several methods exist for calculating fibre. The Southgate, Englyst, and AOAC methods are the most common. Each take different food components into account and so each method can look at the same food and give a different result. The AOAC method is the most commonly used method for showing fibre on food labels in the UK.

Table 7-2	Getting Fibre from Food
Food	*Grams of Fibre in a 100-gram Serving*
Bread	
Brown bread	3.5
Granary bread	4.3
Pitta bread (white)	2.2
White bread	1.5
Wholemeal bread	5.8
Cereals	
Bran cereal	24.5
Bran flakes	13.0
Cornflakes	0.9
Porridge (old-fashioned oats)	9.0
Wheat biscuits	9.7
Grains	
Egg noodles, boiled	0.6
Oatmeal, quick cook, raw	7.1
Rice, brown, cooked	0.8
Rice, white, cooked	0.1
Spaghetti, white, boiled	1.2

(continued)

Table 7-2 *(continued)*

Food	Grams of Fibre in a 100-gram Serving
Spaghetti, wholemeal, boiled	3.5
Wheat bran	36.4
White flour	3.1
Fruits	
Apple, with skin	1.8
Apricots, dried	6.3
Banana	1.1
Figs, dried	6.9
Orange	1.7
Orange juice	0.1
Pear, raw	2.2
Prunes, ready to eat (dried)	5.7
Vegetables (cooked unless stated otherwise)	
Baked beans	3.7
Broccoli	2.3
Brussels sprouts,	3.1
Cabbage	1.8
Carrots	2.5
Cauliflower	1.6
Celery	1.1
New potatoes, boiled in skin	1.5
Peas, frozen	5.1
Potatoes, white, baked with skin	2.7
Red kidney beans, tinned	8.5
Tomatoes, raw	1.0
Nuts and seeds	
Almonds, blanched	7.4
Cashew nuts, roasted and salted	3.2
Peanuts, dry-roasted	6.4
Sunflower seeds	6.0
Walnuts	3.5

Source: Royal Society of Chemistry & Ministry of Agriculture, Fisheries and Foods, McCance & Widdowson's The Composition of Foods 6th Ed, Royal Society of Chemistry, UK, 1994

Getting to the heart of the matter with oat bran

Scientists know that eating foods high in soluble fibre may help lower your cholesterol, although nobody knows exactly why. One thing we do know is that oats keep you fuller for longer and make you less likely to snack on sugary and fatty snacks after you eat them. Oat bran's 'magic wand' factor is claimed to be the soluble fibre *beta-glucans*. Fruits and vegetables (especially dried beans) are also high in soluble fibre as well as heart-friendly antioxidant vitamins and minerals, but ounce for ounce, oats have more soluble fibre. In addition, beta-glucans are thought to be a more effective cholesterol buster than pectin and gum, which are the soluble fibres in most fruits and vegetables.

In 1999 the British Nutrition Foundation summarised the effects of the 'cardio-protective diet', using evidence gathered from three major studies (the DART trial in 1998, the Lyon Diet Heart Study in 1994 and the Indian cardio-protective diet trial in 1992, since you ask). The cardio-protective diet derives from the Mediterranean area where consumption of oily fish, polyunsaturated oils and fruits and vegetables is high. Although the evidence showed the cholesterol-lowering effect of the soluble fibre found in fruits, vegetables and indeed oats to be smaller than had once been thought, the other benefits from soluble fibre-rich foods, such as the antioxidant vitamins and minerals, are sufficiently protective to the heart to justify the message that people should increase consumption.

In the UK it's illegal to claim that a particular food can treat, prevent or cure disease; however, you can imply that a food may benefit health by mentioning how it relates to a disease risk factor. So food manufacturers can state on packaging that 'oats *may* help lower cholesterol when eaten *as part of* a low-fat diet and a healthy lifestyle' (emphasis ours). The crux of the matter is that oats are a healthy, nutritious food, but you can only realise all the benefits of eating oats when they're part of that essential healthy diet and lifestyle.

Oats alone have no magical powers; they're not a cure-all 'superfood'. Food labelling and the types of claims that manufacturers can use on packaging are due to be tightened up in the next few years. We say that these changes are well overdue and can't come soon enough!

Chapter 8

The Truth: The Alcohol Truth

Alcohol is among humankind's oldest pleasures, so highly regarded that the ancient Greeks and Romans called wine a 'gift from the gods', and when whisky was first produced it was named *uisgebeatha* (whis-key-ba), a combination of the Gaelic words for water (*uisge*) and life (*beatha*). Alcohol is unusual in being both a nutrient (providing energy but no other nutritional value) and a drug. Today, although you may enjoy a drink (90 per cent of the UK population consumes alcohol, the majority with no ill effects most of the time), it's important to realise that alcohol has risks as well as a few benefits.

Creating Alcoholic Drinks

When microorganisms (yeasts) digest (ferment) the sugars in food, they make two by-products: a liquid and a gas. The gas is carbon dioxide. The liquid is *ethyl alcohol*, also known as *ethanol*, the intoxicating ingredient in alcoholic drinks.

This biochemical process isn't unusual. In fact, it happens in your own kitchen every time you make bread. Remember the faint, beer-like smell while the dough is rising? That odour is from the alcohol that the yeasts make as they eat their way through the sugars in the flour. (Fortunately, the alcohol evaporates when you bake the bread, otherwise you'd be a bit merry after every sandwich.) As the yeasts digest the sugars, they also produce carbon dioxide, which makes the bread rise.

Whenever you see the word *alcohol* alone in this book, it means ethanol, the only alcohol used in alcoholic drinks.

Alcoholic drinks are produced either through fermentation or through a combination of fermentation plus distillation.

Fermented alcohol products

Fermentation is a simple process in which yeasts or bacteria are added to carbohydrate foods such as corn, potatoes, rice or wheat, which are used as starting material. The yeasts digest the sugars in the food, leaving liquid (alcohol), which is filtered to remove the solids. Water is then added to dilute the alcohol, producing – voilà – an alcoholic beverage.

Distilled alcohol products

The *distillation* process begins with yeasts, which make alcohol from sugars. But yeasts can't thrive in a place where the concentration of alcohol is higher than 20 per cent. To concentrate the alcohol and separate it from the rest of the ingredients in the fermented liquid, distillers pour the fermented liquid into a *still*, a large vat with a wide, column-like tube on top. The still is heated so that the alcohol, which boils at a lower temperature than everything else in the vat, turns to vapour, which rises through the column on top of the still, to be collected in containers where it condenses back into a liquid.

The alcohol you never want to drink

Ethanol is the only kind of alcohol used in food and alcoholic drinks, but it isn't the only kind of alcohol used in consumer products. Here are the other alcohols that may be sitting on the shelf in your bathroom or garage:

✔ **Cosmetic alcohol:** Ethanol is used in cosmetics or medicines such as surgical spirit, which can be used for sterilising the skin. However, it's treated or *denatured* to make it smell and taste awful so that you won't drink it by mistake. Some denaturants (the chemicals used to denature the alcohol) are poisonous.

✔ **Methyl alcohol (methanol):** This poisonous alcohol is made from wood. It's used as a *chemical solvent* (a liquid that dissolves other chemicals) in antifreeze and as a denaturant for ethanol, producing methylated sprits.

✔ **Isopropyl alcohol:** Another poisonous alcohol, made from *propylene*, a petroleum derivative. It includes a substance that makes it taste and smell revolting. It's also used in antifreeze, some lotions and cosmetics, and as a sterilising solution.

This alcohol, called *neutral spirits*, is the base for the alcoholic drinks called spirits or distilled spirits: gin, rum, tequila, whisky and vodka. (Brandy is a special product, a fermented wine that's then distilled. Cognac is distilled grape wine; pear brandy is distilled wine made from pears.)

Foods used to make alcoholic drinks

You can make alcohol from virtually any carbohydrate-containing food. The foods most commonly used are cereal grains, fruit, honey or potatoes. All produce alcohol, but the alcohols have slightly different flavours and colours. Table 8-1 shows you which foods are used to produce the different kinds of alcoholic drinks.

Table 8-1	Which Food Makes What Drink
Start With This Food	*To Get This Alcoholic Drink*
Fruit and fruit juice	
Agave (a type of cactus plant)	Tequila
Apples	Cider
Grapes and other fruits	Wine
Grain	
Barley	Beer, various distilled spirits, kvass
Corn	Bourbon, corn whisky, beer
Rice	Sake (a distilled product), rice wine
Rye	Whisky
Wheat	Distilled spirits, beer
Others	
Honey	Mead
Milk	Kumiss (koumiss)
Potatoes	Some types of vodka
Sugar cane	Rum

Hic! How Much Alcohol is in That Bottle?

No alcoholic drink is 100 per cent alcohol. It's alcohol plus water, and – if it's a wine or beer – some residue of the foods from which it was made.

The label on every bottle of wine and spirits shows the alcohol content as *alcohol by volume* (ABV). ABV measures the amount of alcohol as a percentage of all the liquid in the container. For example, if your container holds 100 millilitres of liquid and 10 millilitres of that is alcohol, the product is 10 per cent ABV. To make it easier to assess intake, you can think of alcohol in terms of units – one unit is equal to 8 grams or 10 millilitres of pure alcohol. The ABV on the label tells you the number of units in 1 litre. So a 12 per cent wine contains 12 units in a litre, or 9 units in a standard 750-millilitre bottle, or 1.5 units per 125-millilitre glass. Labels on most alcoholic drinks in the UK now show the number of units of alcohol in the bottle or can. Table 8-2 gives you a handy at-a-glance guide to the units of alcohol in the most common tipples.

Table 8-2	Units of Alcohol in Common Pub Drinks		
Drink and ABV	*Serving Size*	*Calories*	*Units*
Beer, lager or cider			
3.5% alcohol	1/2 pint (284ml)	90	1
5% alcohol	1/2 pint (284ml)	140	1.5
8–9% alcohol	1/2 pint (284ml)	190	2.5
Wine (12% alcohol)			
Red or dry white	1 small glass (125ml)	70	1.5
Sweet white	1 small glass (125ml)	100	1.5
Spirits (40% alcohol)	25ml pub measure	50	1
Fortified wines (sherry, port or aperitifs: 20% alcohol)	50ml pub measure	60–80	1
Alcopops (fizzy fruit-based drinks containing 5% alcohol)	Bottle (275ml)	180	1.5

Moving Alcohol through Your Body

On its own, alcohol provides energy (7 calories per gram) but no other nutrients. Spirits, such as whisky or vodka, also have no nutrients other than calories. Beer, wine, cider and other fermented beverages contain very small amounts of protein, carbohydrate, vitamins and minerals. Contrary to popular belief, dark beers like stout aren't a particularly good source of iron!

Other nutrients must be digested before being absorbed by your cells, but alcohol gets an Express Check-In – your body doesn't needs to digest alcohol so it absorbs and metabolises the booze before the other nutrients get a look in. In fact, your body absorbs alcohol so fast and so efficiently that about 20 per cent of the alcohol in your drink reaches your brain within a minute of drinking it. Your body absorbs alcohol especially quickly on an empty stomach (food delays it leaving the stomach for the small intestine where most absorption occurs) or from a carbonated drink such as sparkling wine, or whisky and soda.

Your blood carries alcohol to nearly every organ in your body. The following is a road map to show you the route the alcohol travels from every drink you take.

Flowing down the hatch from mouth to stomach

Alcohol is *astringent*: it coagulates the proteins on the surface of the lining of your cheeks and makes them 'pucker'. Some alcohol is absorbed through these tissues and some through the lining of your throat; but most of the alcohol you drink spills into your stomach where further absorption takes place. Here enzymes called *alcohol dehydrogenases* (ADH) begin to break it down. Stomach ADH is responsible for preventing about 20 per cent of the alcohol you drink from entering the blood.

The alcohol dehydrogenases your body produces are influenced by your race and your gender. For example, some ethnic groups such as the Japanese, South Asians, Aborigines and Inuit appear to secrete fewer dehydrogenases than do Caucasians. As a result, they can experience characteristic nausea, headache and flushing when they drink even small amounts of alcohol. Similarly, for the average woman ADH are up to 40 per cent less active than in men – as a result they're likely to become drunk on smaller amounts of alcohol than it takes to send an average male under the table. So don't match drink for drink, ladies!

Stopping for a short visit at the energy factory

Alcohol that survives the stomach travels on down to the small intestine from where you absorb it via a large blood vessel (the *portal vein*) and carry it to the liver. In the liver, another group of alcohol dehygrogenases metabolise the remaining alcohol to produce energy using a coenzyme called *nicotinamide adenine dinucleotide* (NAD). Try saying that after a couple of drinks!. A normal, healthy liver can only process about one unit of alcohol an hour in this way.

While you're body is using NAD for alcohol, it can't use it to run your other metabolic pathways. As a result these become blocked, causing problems such as the build-up of fat in the liver and eventual liver damage.

Rising to the surface

While waiting to be metabolised, alcohol may travel around the body in the blood. Alcohol can make blood *platelets* (tiny particles that enable blood to clot) less sticky and make *fibrinogen*, a natural blood clotting agent, less effective, which may be one reason why moderate amounts of alcohol help protect against heart disease (see the later section 'Alcohol and Health').

Alcohol makes blood vessels expand, so more warm blood flows up from the centre of your body to the surface of the skin. You feel warmer for a while and, if your skin is fair, you may flush and turn pink.

Encountering curves in the road

Alcohol is a sedative. When it crosses from your blood into your brain, it slows the transmission of impulses between nerve cells that control your ability to think and move. That's why your thinking may be fuzzy, your judgement impaired, your tongue twisted, your vision blurred.

Why do you feel the urge to pee more when you drink alcohol? It's not just the added volume of fluid; alcohol reduces your brain's production of *antidiuretic hormones*, which keep you from making too much urine. You may lose lots of fluid and electrolytes by peeing more (see Chapter 12). You also grow very thirsty as you become more and more dehydrated, and your urine may smell faintly of alcohol. This cycle continues as long as you have alcohol circulating in your blood, or in other words, until your liver manages to produce enough ADH to metabolise all the alcohol you've consumed. How long is that? Most people need a full hour to metabolise the amount of alcohol in one unit. But that's only an average: some people still have enough alcohol circulating in their blood to impair normal functioning for as long as two to three hours after they take a drink.

Alcohol and Health

Alcohol has benefits as well as side effects. The benefits seem to be linked to what the Department of Health calls 'sensible drinking guidelines', based on the effects of alcohol at different levels of intake. Initially, these guidelines were expressed as units per week (no more than 14 units for women and 21 units for men), but in recognition of the risks of *binge drinking* (excess drinking in one session), daily guidelines now exist. Currently, these daily guidelines state that women shouldn't regularly drink more than two to three units a day and men shouldn't regularly drink more than three to four units a day. In addition, you should spread units in the week, avoiding binge drinking. Unfortunately the recent Office for National Statistics General Household Survey showed that in the UK around 41 per cent of men and 34 per cent of women report that they regularly exceed these safe levels.

Moderate drinking: Some benefits, some risks

Moderate amounts of alcohol reduce stress, so it isn't surprising that recent well-designed scientific studies on large groups of men and women suggest that moderate drinking is heart healthy, protecting the cardiovascular system. Other benefits include:

- An intake of one to two units a day can have a moderately protective effect against heart disease. This effect may be because it increases 'good' cholesterol (see Chapter 5) or from the reductive effects on blood clotting. However, men over the age of 40 and post-menopausal women appear to benefit most from moderate drinking.

- People who drink moderately are less likely to suffer a blood clot-related (*ischaemic*) stroke, which accounts for around 80 per cent of all strokes in the UK. One meta-analysis (a study analysing other studies) based on 35 studies of alcohol and stroke showed that men who took one to two drinks a day were 30 per cent less likely than those who never drank to have a clot-related stroke. However, the downside is that even moderate alcohol increases the risk of another less common type of stroke resulting from a burst blood vessel in the brain: *haemorrhagic* stroke, and heavy alcohol intake dramatically increases the incidence of both types of stroke.

✔ Moderate drinking may even help protect the brain into old age. A recent study carried out in Spain and published in the *Journal of Alzheimer's Disease* showed that those who drank moderately (especially women and non smokers) were less likely to suffer from Alzheimer's disease later in life. However, drinking too much has an adverse effect: other studies have shown heavy drinkers to be more at risk of cognitive decline and dementia.

A small margin exists between benefit and risk. Harm starts to outweigh benefit after you exceed the sensible levels of alcohol consumption and increases dramatically in intakes over 40 units per week. Indeed, excessive drinking is as harmful as tobacco smoking. High alcohol is linked to over 50 diseases, including several types of cancer – especially head and neck but also possibly liver, pancreas, rectum and breast.

High alcohol intakes raise blood pressure and excess consumption can cause resistance to treatment for high blood pressure. Regular heavy drinkers may find it difficult to control their weight because alcohol is packed with empty calories (7 calories per gram). Recent studies suggest that alcohol may even be a risk factor for osteoporosis or brittle bone disease. Alcohol also affects the brain and can act as a depressant, and it can irritate the oesophagus and cause stomach inflammation.

Tips for cutting down your alcohol intake

✔ If you're thirsty, quench your thirst with water or another soft drink before you start on alcohol.

✔ Alternate alcoholic drinks with nonalcoholic drinks or drink a glass of water alongside.

✔ Choose weaker beers and wines – those with a lower ABV – or dilute well with non-alcoholic mixers such as soda, lemonade or tonic water.

✔ If pouring spirits at home go easy on the measures or use a pub measure to guide you.

✔ Try to avoid buying drinks in rounds if this means you drink faster or more than you intend to, or choose half pints instead of pints.

✔ Avoid eating salty foods such as crisps and nuts because this will make you drink more – bartenders often dish these out for reason, you know!

My head hurts! Having a hangover

Excessive drinking can make you feel terrible the next day. 'The morning after' is no joke. A hangover is a miserable physical fact:

- ✔ You're thirsty because you lost excess water through copious urination.

- ✔ Your stomach hurts and you're nauseous because even small amounts of alcohol irritate your stomach lining, causing it to secrete extra acid and lots of *histamine*, the same immune system chemical that makes the skin around a mosquito bite red and itchy.

- ✔ Your muscles ache and your head pounds because processing alcohol through your liver requires a co-enzyme – nicotinamide adenine dinucleotide, remember? NAD is normally used in alternative metabolic pathways such as converting *lactic acid*, a by-product of muscle activity, to other chemicals that you can use for energy. The extra, unprocessed lactic acid piles up painfully in your muscles.

When you've overdone it, take a complete break from alcohol for 48 hours to let your body recover.

Alcoholism and nutrition

Alcohol isn't only a nutrient, it's a drug, albeit a legal and socially accepted one. Alcohol activates the pleasure and reward centres of the brain by stimulating the production of neurotransmitters. However, some people go overboard with their drinking. *Alcohol abuse* is a term generally taken to mean drinking so much that it interferes with your ability to have a normal, productive life. The costs of excessive drinking, both to the health of the individual and UK society, are well documented. Untreated alcohol abuse is a life-threatening disease that can lead to death either from an accident or suicide (both are more common among heavy drinkers); from a toxic reaction (acute alcohol poisoning that paralyses body organs, including heart and lungs); or malnutrition or liver damage (fatty liver and cirrhosis). In the UK over 27,000 people die from alcohol-related illnesses each year, costing the NHS around £2.7 billion annually.

Alcoholism makes it extremely difficult for the body to get essential nutrients. Here's why:

- ✔ Alcohol depresses appetite, causing a loss of interest in food, and leads to gastric irritation – all reducing food intake.

- ✔ An alcoholic may substitute alcohol for food, getting calories but no nutrients.

✔ Even when the alcoholic eats, the presence of alcohol in his or her tissues can deplete the body of vitamins (notably the B vitamins), minerals and other nutrients. It also reduces the alcoholic's ability to synthesise protein and raises blood fat (triglycerides) levels.

No one knows exactly why some people are able to have a drink once a day or once a month or once a year, enjoy it and move on, while others become addicted to alcohol. In the past, alcoholism has been blamed on heredity (bad genes), lack of willpower or even a bad upbringing. But as science continues to unravel the mysteries of body chemistry, it's reasonable to expect that researchers will eventually come up with a rational scientific explanation for the differences between social drinkers and people who can't safely use alcohol. It just hasn't happened yet.

If you think you or a friend or relative may have a problem with alcohol, contact your GP or Drinkline (freephone 0800 917 8282), which can provide confidential information and advice and put you in touch with local alcohol support services.

Avoiding alcohol

No one should drink to excess. But some people should be especially careful.

For example:

✔ **People who plan to drive or to do work that requires both attention and skill.** Alcohol slows reaction time and makes your motor skills – turning the wheel of the car, operating heavy machinery – less precise. In simulated driving tests, bus drivers with blood alcohol levels well under the legal limit for driving thought they could drive through obstacles that were too narrow for their vehicles. Even at the legal limit they were twice as likely to encounter road traffic incidents. An estimated one in five road deaths and as many as one in four accidents in the workplace are related to alcohol.

✔ **Women who are pregnant or who plan to become pregnant in the near future.** Alcohol can make conception more difficult and it can also pass through the placenta into the foetal bloodstream increasing the chances of miscarriage, especially in the first three months of pregnancy. Heavy drinking during pregnancy increases the risk of a baby being born with a disorder called *foetal alcohol syndrome* (FAS). This is a collection of birth defects including low birth weight, heart defects, retardation and facial deformities documented only in babies born to heavy drinkers. Current Department of Health advice is that pregnant women or women trying to conceive should avoid drinking alcohol.

If they do choose to drink they should drink no more than one to two units of alcohol once or twice a week. Drinking even three units per week increases miscarriage rates.

Women are generally more susceptible to damage from alcohol because their bodies contain less water to dilute the alcohol. Women can tolerate less alcohol before damage occurs to organs such as the liver, pancreas, brain and heart.

✔ **People who are currently taking prescription drugs or over-the-counter medication.** Alcohol makes some drugs stronger, increases the side effects and renders others less effective. At the same time, some drugs make alcohol a more powerful sedative or slow down the elimination of alcohol from your body. In people with diabetes treated with insulin or tablets alcohol can cause the blood sugar to go too low (hypoglycaemia). If you take any kind of medication, check with your doctor or pharmacist regarding the possibility of an interaction with alcohol.

Good advice endures the test of time. Even the Bible says: 'Wine is as good as life to man if it be drunk moderately.' We'd add to that: 'Keep to the safe levels and avoid binge drinking!' And it's impossible to improve on this slogan from the Romans (actually, one Roman writer named Terence): 'Moderation in all things.' You can't get a message more sensible than that.

The power of purple

Grape skin contains *resveratrol*, a naturally occurring plant chemical. The darker the grapes, the higher the concentration of resveratrol (red wine has more resveratrol than white wine and red grape juice has more resveratrol than white grape juice).

Some population studies have suggested that resveratrol may help to reduce the risk of heart disease and some kinds of cancer, which may help explain some of the benefits of moderate drinking. But before you start glugging red wine, remember that this idea hasn't yet been tested in clinical trials in humans, so the jury is still out on the safety and benefits of taking extra resveratrol. Our best advice remains to eat a balanced diet that contains a variety of colours of fruits and vegetables, and to drink in moderation only.

Chapter 9

Vigorous Vitamins

● ●

In This Chapter

▶ Understanding the value of vitamins

▶ Revealing the best food sources for the vitamins you need

▶ Discovering the consequences of taking too many (or too few) vitamins

▶ Knowing when you may need extra vitamins

● ●

*V*itamins are *organic chemicals*, substances that contain carbon, hydrogen and oxygen. They occur naturally in all living things, plants and animals alike: flowers, trees, fruits, vegetables, chickens, fish, cows – and you.

Vitamins regulate a variety of bodily functions. They're essential for building body tissues such as bones, skin, glands, nerves and blood. They assist in metabolising (digesting) proteins, fats and carbohydrates, so that you can get energy from food. They prevent nutritional deficiency diseases, promote healing and encourage good health.

This chapter is a guide to where you can find vitamins, how you add them to your diet and how to tell whether you're taking too many vitamins.

Taking a Look at the Vitamins Your Body Needs

Your body needs at least 11 specific vitamins: vitamin A, vitamin D, vitamin E, vitamin K, vitamin C, and the members of the B vitamin family: thiamin (vitamin B1), riboflavin (B2), niacin, vitamin B6, folate and vitamin B12. Two more B vitamins, biotin and pantothenic acid, are also valuable to your wellbeing. You need only minuscule quantities of vitamins for good health. In some cases, the reference nutrient intake (RNI) determined by the Department of Health as the amount required by virtually every healthy person may be as small as just a few micrograms (µgs) or milligrams (mgs).

Introducing Casimir Funk, father of vitamins

Vitamins are so much a part of modern life that you may have a hard time believing they were first discovered fewer than 90 years ago.

People have long known that certain foods contain something special.

The ancient Greek physician Hippocrates prescribed liver for *night-blindness* (the inability to see well in dim light). By 1795 British Navy ships carried a mandatory supply of limes or lime juice to prevent scurvy among the men (thus earning the Brits once and forever the nickname Limies). Later on, the Japanese Navy gave its sailors wholegrain barley to ward off beriberi.

Everyone knew these prescriptions worked, but nobody knew why until 1912, when Casimir Funk (1884–1967), a Polish biochemist working first in England and then in the United States, came up with the theory that the 'somethings' in food were substances that he called *vitamines* (*vita* = life; *amines* = nitrogen compounds).

The following year, Funk and a British biochemist, Frederick Hopkins, proposed a second theory: conditions such as scurvy and beriberi were simply deficiency diseases caused by the absence of a specific nutrient in the body. Adding a food with the missing nutrient to one's diet would prevent or cure the deficiency disease.

Eureka!

Nutritionists classify vitamins as either *fat soluble* or *water soluble*, meaning that they dissolve either in fat or in water. If you consume larger amounts of fat-soluble vitamins than your body needs, the excess is stored in body fat. Excess water-soluble vitamins are eliminated in urine.

Vitamins are often classified according to their function within specific body tissues, such as folate and vitamin B12 developing and maintaining blood, or vitamin D maintaining bone health.

Another classification of vitamins relates to their effect on general body systems, such as the antioxidant function of vitamins C and E.

Whichever way you choose to classify your vitamins, the bottom line is that most of them are essential: your body either can't make them at all or can't do so in sufficient quantities, so you rely on food or supplements to give you your daily dose. Large amounts of fat-soluble vitamins stored in your body may cause problems (see the section 'Fat-soluble vitamins' in this chapter). With water-soluble vitamins, your body simply shrugs its shoulders and urinates away most of the excess.

Use this mnemonic to remember which vitamins are fat soluble and which dissolve in water: '**A**ll **D**ogs **E**at **K**idneys.' Vitamins A, D, E and K are fat soluble. All the rest dissolve in water.

Fat-soluble vitamins

Vitamin A, vitamin D, vitamin E and vitamin K are relatives that have two characteristics in common: all dissolve in fat, and all are stored in your fatty tissues. But like members of any family, they also have distinct personalities. One keeps your skin hydrated. Another protects your bones. A third keeps reproductive organs purring happily. And the fourth enables you to make special proteins. Because you can store fat-soluble vitamins, you could say that they work a bit more like a savings account where you can dip into your stash when you need it, whereas the water-soluble vitamins work more like your current account and need topping up regularly. For instance, the fat-soluble vitamin D is made in the skin from sunlight – you tend to build up stores in the summer but can still use those stores in the long, dark, damp winter months. Clever, eh?

Vitamin A

Vitamin A is the moisturising nutrient. This vitamin keeps your skin and *mucous membranes* (the smooth tissue that lines the eyes, nose, mouth, throat, vagina and rectum) smooth and supple. Vitamin A is also the vision vitamin, producing *rhodopsin* in the retina, making it possible to see when the lights are low. Vitamin A also promotes the growth of healthy bones and teeth, keeps your reproductive system healthy and encourages your immune system to churn out the cells you need to fight off infection.

Two chemicals provide vitamin A: retinoids and carotenoids. *Retinoids* are compounds whose names all start with *ret*: retinol, retinaldehyde, retinoic acid and so on. These fat-soluble substances are found in several foods of animal origin – liver, whole milk, eggs and butter and other fat spreads. Retinoids give you *preformed* vitamin A, the kind of nutrient your body can use right away. The second form of vitamin A is some of the *carotenoids*: yellow, red and dark green pigments in fruits and vegetables. Carotenoids are vitamin A *precursors*, chemicals your body transforms into retinol-like substances.

So far, scientists have identified at least 500 different carotenoids. Only about 50 are considered to be sources of vitamin A. The star of the lot is *beta-carotene*, a pigment found in most bright yellow and deep green fruits and vegetables.

Carotenoids play an important role in vision, helping to prevent or slow the development of age-related *macular degeneration*, progressive damage to the retina of the eye, which can cause the loss of central vision (the ability to see clearly enough to read or do fine work).

The RNI of vitamin A for children aged 1–10 years is 400–500 micrograms per day rising to 600 per day for 11–14 year olds. Over this age you need 700 micrograms per day for men and 600 for women, but this increases during pregnancy to 700 micrograms per day and to 950 while breast-feeding.

Help! I'm turning orange

Carotenoids (a form of vitamin A) are stored in body fat. If you wolf down large quantities of carotenoid-rich foods like carrots and tomatoes every day, day after day, for several weeks, your skin – particularly the palms of your hands and the soles of your feet – turns a nifty shade of dusty orange, brighter if your skin is naturally light, darker if it's naturally dark. Cut down on the carrots and tomatoes, and the orange colour fades.

People living in the UK tend to eat lots of meat, dairy products and fat spreads, so that many far exceed typical daily vitamin A requirements. Deficiency is rare and generally only seen in people who can't absorb fat or who follow very low-fat diets. Very high doses of vitamin A can cause damage to the liver, so supplements aren't a good idea. Pregnant women are advised not to eat liver because it's such a rich vitamin A source – you store vitamin A in the liver, and so do the animals you eat. Women who have been through the menopause and older men who are more at risk of osteoporosis should avoid having more than 1.5mg of vitamin A a day.

Vitamin D

If we say 'bones' or 'teeth', what nutrient springs most quickly to mind? If you answer calcium, you're only partly right. True, calcium is essential for hardening teeth and bones. But no matter how much calcium you consume, without vitamin D your body can't absorb and use calcium. As well as it being vital for strong bones and teeth, some evidence points to the fact that failing to get enough vitamin D may increase the risks of diabetes and some forms of cancer.

<NutritionSpeak>Vitamin D comes in three forms: calciferol, cholecalciferol (vitamin D3) and ergocalciferol (vitamin D2). *Calciferol* occurs naturally in fish oils and egg yolk. In the United Kingdom it's mandatorily added to fat spreads and voluntarily added to some breakfast cereals and some dairy products. *Cholecalciferol* is created when sunlight hits your skin and ultraviolet rays react with steroid chemicals in body fat just underneath. Sunlight is your main source of vitamin D, despite Britain's famously grey weather. *Ergocalciferol* is synthesised in plants exposed to sunlight. Cholecalciferol and ergocalciferol justify vitamin D's nickname: the sunshine vitamin. Both forms are biologically inactive when consumed in foods until they've been converted into the active form by the liver and the kidneys.

The RNI for vitamin D is measured in micrograms (µg). For infants under 6 months the RNI is 8.5 micrograms per day and then up to 4 years is 7 micrograms per day. People aged 4–64 are assumed not to need a dietary source as long as they have some exposure to sunlight during spring and summer.

Risk of vitamin D deficiency is low in the UK, but may be higher among elderly people, especially those in care homes or the housebound. Some ethnic groups that require complete skin covering with clothes are also at risk. Prolonged deficiency in children results in rickets, and osteomalcia and osteoporosis in adults. Very high dietary intakes of vitamin D can result in *hyper* (high) *vitaminosis D*, which can result in dangerously high levels of calcium entering the blood, but you won't get it from too much sunbathing!

Vitamin E

Every animal, including you, needs vitamin E to maintain a healthy reproductive system, nerves and muscles. Vitamin E is also an important antioxidant and because it's fat soluble, it's involved in protecting every body cell from the effects of *free radicals* (unstable compounds that damage healthy body cells). You get vitamin E from *tocopherols* and *tocotrieonols*, two families of naturally occurring chemicals in vegetable oils (particularly sunflower oil and wheatgerm oil), nuts, whole grains and green leafy vegetables. These are your best natural sources of vitamin E. Tocopherols, the more important source, have two sterling characteristics: they're anticoagulants and antioxidants.

Hand in hand: How vitamins help each other

All vitamins have specific jobs in your body. Some have partners. Here are some examples of nutrient cooperation:

- Vitamin E keeps vitamin A from being destroyed in your intestines.

- Vitamin D enables your body to absorb calcium and phosphorus.

- Vitamin C helps folate build proteins.

- Vitamin B1 works in digestive enzyme systems with niacin, pantothenic acid and magnesium.

This is another reason nutritionists recommend that if you do choose to take a vitamin supplement you should take one good-quality multivitamin and mineral supplement rather than a cocktail of single vitamins.

In addition, taking vitamins with other vitamins may improve body levels of nutrients. In 1993 scientists at the National Cancer Institute and the United States Department of Agriculture (USDA) Agricultural Research Service gave one group of volunteers a vitamin E capsule plus a multivitamin pill; a second group, the same amount of vitamin E alone; and a third group, no vitamins at all. The people getting vitamin E plus the multivitamin had the highest amount of vitamin E in their blood – more than twice as high as those who took plain vitamin E capsules.

Sometimes, one vitamin may even alleviate a deficiency caused by the lack of another vitamin. People who don't get enough folate are at risk of a form of anaemia in which their red blood cells fail to mature. As soon as they receive folate, either by injection or by mouth, they begin making new healthy cells. That's to be expected. What's surprising is the fact that anaemia caused by *pellegra,* the niacin deficiency disease, may also respond to folate treatment.

Isn't nature clever?

  ✔ *Anticoagulants* reduce the blood's ability to clot and may reduce the risk
    of clot-related stroke and heart attack.

  ✔ *Antioxidants* prevent free radicals (incomplete pieces of molecules) from
    joining up with other molecules or fragments of molecules to form toxic
    substances that can attack tissues in your body.

Can tocopherols lower cholesterol? Could be. Tocopherols ride around your
body on the backs of *lipoproteins*, the fat and protein particles that carry cholesterol either into arteries or out of the body. Some research suggests that the
tocopherol passengers may prevent low-density lipoproteins (LDLs, the *bad*
cholesterol) from sticking on artery walls and blocking blood vessels – making
vitamin E great for a healthy heart. (For more on cholesterol, see Chapter 5.)

Vitamin E is fab for heart health too:

  ✔ A clinical trial at Cambridge University shows that taking 800 IU
    (International Units) of vitamin E, twice the RNI, may reduce the risk of
    nonfatal heart attacks for people who already have heart disease.

  ✔ Researchers at the University of Minnesota say that postmenopausal
    women who get at least 10 IU of vitamin E a day from food cut their risk
    of heart disease by nearly 66 per cent.

But don't overdose on vitamin E! The latest news is that too much vitamin E
is potentially hazardous, especially for people who are taking drugs such as
aspirin, which reduces blood clotting. The anticoagulant properties of the
aspirin and vitamin E may compete and reduce the overall effect. However,
it's very unlikely you'd consume enough vitamin E to cause this effect without taking a lot of very concentrated vitamin E supplements.

Because vitamin E does so much work, it's very difficult to calculate an RNI.
Deficiency of vitamin E is so rare it's only really seen in premature babies
or with people who have exceptionally high intakes of polyunsaturated fats
(well above normal healthy consumption). Likewise, toxicity of vitamin E is
not thought to be a risk to humans.

Vitamin K

Vitamin K is a group of chemicals that your body uses to make specialised
proteins found in blood *plasma* (the clear fluid in blood), such as prothrom-
bin, the protein chiefly responsible for blood clotting and stopping you from
bleeding to death when you cut yourself. You also need vitamin K to make
bone and kidney tissues. Like vitamin D, vitamin K is essential for healthy
bones. Vitamin D increases calcium absorption; vitamin K activates at least
three different proteins that take part in forming new bone cells. A report
on 888 men and women from the Framingham (Massachusetts) Heart Study
shows that those who consumed the least vitamin K each day had the highest incidence of broken bones. The same was true for a 1999 analysis of data
from the Nurses' Health Study in the United States.

You get Vitamin K from dark green leafy vegetables (broccoli, cabbage, kale, lettuce, spinach and turnip greens), cheese, liver, cereals and fruits, but most of what you need comes from resident colonies of friendly bacteria in your intestines, an assembly line of busy bugs churning out the vitamin day and night.

As with vitamins D and E, no recommended daily intake exists. The only group at risk of vitamin K deficiency is newborn babies who don't have enough friendly gut bacteria to make their own. In the UK all newborn babies receive an injection of synthetic vitamin K that keeps them going until they can be self-sufficient. Both deficiency and toxicity are very rare.

Water-soluble vitamins

Vitamin C and the entire team of B vitamins (thiamin, riboflavin, niacin, vitamin B6, folate, biotin, pantothenic acid) are usually grouped together simply because they all dissolve in water.

Because these vitamins dissolve in water, your body can't store large amounts of these nutrients. If you take in more than you need to perform specific body tasks, you simply pee away virtually all the excess. The good news is that these vitamins rarely cause side effects. The bad news is that you have to take enough of these vitamins every day to protect yourself against deficiencies.

Vitamin C

Vitamin C, also called *ascorbic acid*, is essential for the development and maintenance of connective tissue (the fat, muscle and bone framework of the human body). Vitamin C speeds up the production of new cells in wound healing; helps with absorption of iron; and, like vitamin E, it's an antioxidant that keeps free radicals from joining up with other molecules to form damaging compounds that may otherwise attack your tissues. Vitamin C protects your immune system, helps you fight off infection and may reduce the severity of allergic reactions. It plays a role in the synthesis of hormones and other body chemicals.

The RNI for children is 30 milligrams per day; 40 milligrams for adults. Increase this dose by 10 milligrams per day if you're pregnant or breastfeeding. Smokers increase their turnover of vitamin C and should double their intake to 80 milligrams per day.

The main sources of vitamin C are vegetables, fruit and fruit juices. The most concentrated sources are citrus fruit, but surprisingly you get around 16 per cent of your daily dose from the humble potato – not so much because potatoes are such a good source, but because us Brits typically eat so many of them. Vitamin C lies mainly beneath the potato's skin, so eating new potatoes with the skins on or using a peeler instead of a knife helps keep your vitamin C levels up.

Thiamin (vitamin B1)

This sulphur (*thia*) and nitrogen (*amin*) compound helps ensure a healthy appetite. It acts as a *coenzyme* (a substance that works along with other enzymes) essential to at least four different processes by which your body metabolises energy from carbohydrates, fats and alcohol. Thiamin also is a mild diuretic (something that makes you urinate more).

Although thiamin is found in every body tissue it's not stored in the body, so you need a daily intake. The RNI is calculated on the basis of energy (calorie) intake at 0.4 milligrams per 100 calories, which in practice levels out to children needing 0.5–0.7 milligrams per day; adult men 1.0 milligrams per day; and adult women 0.8 milligrams per day.

The richest dietary sources of thiamin are unrefined cereals and grains, lean pork, pulses, nuts and seeds. In the UK all flour (with the exception of wholemeal flour) is fortified with thiamin and most breakfast cereals are voluntarily fortified. Thiamin deficiency is rare and most likely to be seen in alcoholics (when the body is so busy dealing with alcohol it's less effective at absorbing nutrients) as Wernicke-Korsakoff syndrome (a form of dementia).

Riboflavin (vitamin B2)

Like thiamin, riboflavin is a coenzyme. Without it, your body can't digest and use proteins and carbohydrates. Like vitamin A, it protects the health of mucous membranes. You get riboflavin from foods of animal origin (meat, fish, poultry, eggs and milk), whole or enriched grain products, brewer's yeast and dark green vegetables such as broccoli and spinach.

The RNIs for riboflavin are 0.6–1.0 milligrams per day for children, 1.3 milligrams per day for men and 1.1 milligrams per day for women, with pregnant and breast-feeding women requiring a little more. Most children and adults exceed their needs for riboflavin in their normal diet, but up to 20 per cent of older people have intakes below the RNI because older people consume less riboflavin rich foods and don't absorb nutrients as well as younger people. Deficiencies show as cracks around the mouth and nose, which are quickly relieved once intake goes up again. It's highly unlikely for you to reach toxic levels of riboflavin because the body is extremely efficient at eliminating excess riboflavin through urine.

Niacin (Vitamin B3)

Niacin is one name for a pair of naturally occurring nutrients, nicotinic acid and nicotinamide. Niacin is essential for proper growth, and like other B vitamins it's intimately involved in enzyme reactions. In fact, niacin is an integral part of an enzyme that enables oxygen to flow into body tissues. Like thiamin, it gives you a healthy appetite and participates in the metabolism of sugars and fats.

Atishoo! Vitamin C and the common cold

Does taking vitamin C supplements prevent colds and flu, as so many people believe? No evidence exists that vitamin C supplements can prevent colds – although they may reduce the length of time you have to put up with the cold symptoms. Higher doses of 100 milligrams per day produce better benefits than small doses.

Before you start downing vitamin C capsules like jelly beans, high doses of more than 1 gram a day can result in diarrhoea and lead to formation of kidney stones. Our advice: if you take high-dose vitamin C supplements, don't do it long term; wait until your cold starts before you start taking supplements and then reduce the dose as you begin to feel better. Keep track of how much you take: more than 1 gram a day might do you more harm than good.

Niacin is available either as a preformed nutrient or via the conversion of the amino acid tryptophane. Preformed niacin comes from meat; tryptophane comes from milk and dairy foods. Some niacin is present in grains, but your body can't absorb it efficiently unless the grain has been treated with lime (the mineral, not the fruit). This is a common practice in Central and South American countries where lime is added to cornmeal in making tortillas. In the United States and in the United Kingdom, breads and cereals are routinely fortified with niacin. Your body easily absorbs added niacin.

Like thiamin, the RNIs for niacin relate to energy (calorie) intake, with 6.6 milligrams of niacin per 100 calories. For children that means 8–12 milligrams per day, for adult men 17 milligrams per day and adult women 13 milligrams per day. In practice, even the lowest recorded intakes of niacin were still above the RNIs, so the niacin deficiency, pellagra, is virtually unheard of in the UK. Very high intakes – again, way above normal – can have adverse effects on the liver.

Vitamin B6 (Pyridoxine)

Vitamin B6 is another multiple compound, this one comprising three related chemicals: pyridoxine, pyridoxal and pyridoxamine. Vitamin B6, a component of enzymes that metabolises proteins and fats, is essential for getting energy and nutrients from food. It plays an important role in removing excess amounts of homocysteine (see Chapter 4) from your blood. A high level of *homocysteine*, an amino acid produced when you digest proteins, is an independent risk factor for heart disease, perhaps as important as your cholesterol levels.

The best food sources of vitamin B6 are liver, chicken, fish, pork, lamb, milk, eggs, brown rice, whole grains, beer, soya beans, potatoes, beans, nuts, seeds and dark green vegetables such as turnip greens. The RNIs for vitamin B6 are related to protein intake and although deficiencies are rare, people following a high-protein diet can be deficient. For the record the RNIs are 0.7–1.0 milligrams per day for children, adult men 1.4 milligrams per day and adult women 1.2 milligrams per day.

Most people get all their vitamin B6 from a healthy and varied diet.

Folate

Folate is an essential nutrient for human beings and other *vertebrates* (animals with backbones). Folate takes part in the synthesis of DNA, the metabolism of proteins and the subsequent synthesis of amino acids used to produce new body cells and tissues. Folate is vital for normal growth and wound healing. An adequate supply of the vitamin is essential for pregnant women to enable them to create new maternal tissue as well as fetal tissue. In addition, an adequate supply of folate dramatically reduces the risk of neural tube defects (such as spina bifida).

Beans, dark green leafy vegetables, liver, yeast and various fruits are excellent food sources of folate, and all multivitamin supplements must now provide 400 micrograms of folate per dose. The RNIs for folate are 70–150 micrograms (µg) per day for children and 200 micrograms per day for adults. During pregnancy this increases by 100 micrograms per day and by 60 micrograms per day while breast-feeding. Women who intend to become pregnant can consume an additional 400 micrograms per day. This level is difficult to achieve from diet alone, so folic acid is one of the few vitamins where supplements are strongly recommended before conception and during the first 12 weeks of pregnancy. High intakes of folate have no adverse effects.

Deficiency of folate results in megaloblastic anaemia, but rather than being a result of a low intake, deficiency is more attributable to poor absorption or to drug interactions, particularly if you take anti-convulsant drugs for epilepsy.

Vitamin B12 (Cyanocobalamin)

Vitamin B12 makes healthy red blood cells. It protects *myelin*, the fatty material that covers your nerves and enables you to transmit electrical impulses (messages) between nerve cells. These messages make it possible for you to see, hear, think, move and do all the things a healthy body does each day.

Vitamin B6 supplements and PMS: Fact or fiction?

Many women take vitamin B6 supplements for the relief of PMS symptoms, but the supporting scientific evidence is patchy to say the least. The dosage of supplements varies enormously too, from 200 milligrams to 800 milligrams per day. The higher end of this range can lead to *sensory neuropathy* – that's numbness, weakness and pins and needles to you! The Food Standards Agency recommends that supplements should contain no more than 10 milligrams of vitamin B6, but this is only a voluntary guideline. In the United States the upper guideline for supplements is 100 milligrams. The debate goes on.

Vitamin B12 is unique. First, it's the only vitamin that contains a mineral, cobalt. (Cyanocobalamin, a cobalt compound, is commonly used as vitamin B12 in vitamin pills and nutritional supplements.) Second, it's a vitamin that plants can't make. Like vitamin K, vitamin B12 is made by beneficial bacteria living in your small intestine. Meat, fish, poultry, milk products and eggs are good sources of vitamin B12. Grains don't naturally contain vitamin B12, but like other B vitamins, it's added to many breakfast cereals in the UK. It's one of the few water-soluble vitamins that the liver can store.

The RNIs for vitamin B12 are 0.5 micrograms per day for children and 1.5 per day for adults. Dietary deficiency is rare, and usually only seen in young children following a vegan or macrobiotic diet that excludes fortified breakfast cereals or supplements. Prolonged deficiency of B12 causes pernicious anaemia that, if untreated, can result in irreversible neurological damage and megaloblastic anaemia similar to that seen in folate deficiency. Treatment for deficiency is usually in the form of supplements or injections, because sufferers tend not to absorb dietary vitamin B12 well.

Biotin

Biotin is a B vitamin, a component of enzymes that ferry carbon and oxygen atoms between cells. Biotin helps you metabolise fats and carbohydrates and is essential for synthesising fatty acids and amino acids needed for healthy growth. And biotin seems to prevent a build-up of fat deposits that may interfere with the proper functioning of liver and kidneys. (No, biotin won't keep fat from settling in more visible places, such as your hips.)

Your best food sources of biotin are liver, egg yolk, yeast, nuts and beans. If your diet doesn't give you all the biotin you need, bacteria in your gut synthesises enough to make up the difference. No RNIs for biotin exist, but intakes of between 10–200 micrograms a day are thought to be adequate and safe.

Pantothenic acid

Pantothenic acid, another B vitamin, is vital to enzyme reactions that make it possible for you to use carbohydrates and create steroid biochemicals such as hormones. Pantothenic acid also helps stabilise blood sugar levels, defends against infection and protects *haemoglobin* (the protein in red blood cells that carries oxygen through the body), as well as nerve, brain and muscle tissue. You get pantothenic acid from meat, fish and poultry, beans, wholegrain cereals and fortified grain products. There are no RNIs for pantothenic acid, but the current average intake in the UK of 3–7 milligrams a day is thought to be adequate. Large doses may cause diarrhoea and stomach upset.

Get Your Vitamins Here

Reference nutrient intakes present safe and effective amounts of vitamins for healthy people. Head to the cheat sheet for an at-a-glance table of the RNIs for vitamins.

The easier and common-sense option is to keep it simple and eat a well-balanced, varied diet from all the food groups, including your five portions of different-coloured fruit and veg each day. Eating sensibly means you have all the vitamins you need for a long, healthy life – no rocket science, no hassle.

If you want to back up your diet with a good-quality multivitamin and mineral supplement, that's fine. Go for a once-daily tablet that just meets the RNIs for all your nutrients. However, a supplement is only an insurance policy and can never replace the nutritional wonders of our favourite topic – a healthy diet.

Too Much or Too Little: Avoiding Two Ways to Go Wrong with Vitamins

Reference nutrient intakes (RNIs) are broad enough to prevent vitamin deficiencies in most people and avoid the side effects associated with very large doses of some vitamins. If your diet doesn't meet these guidelines, or if you take very large amounts of vitamins as supplements, you may be in for trouble.

Vitamin deficiencies

The good news is that vitamin deficiencies are rare among people who have access to a wide variety of foods and know how to put together a balanced diet. For example, the only people likely to experience a vitamin E deficiency are premature or low-birthweight infants and people with a metabolic disorder that keeps them from absorbing fat. A healthy adult may go as long as ten years on a vitamin E-deficient diet without developing any signs of a problem.

You can increase the quality of the vitamins you get from food. Stale food, particularly fruit and vegetables that have been exposed to air, heat or light, are likely to be less rich in some vitamins than really fresh, well-stored produce (a dark, dry, cool environment such as a cupboard or larder). Frozen vegetables often have as many, if not more vitamins than some fresh varieties simply because they're harvested and frozen so quickly that the vitamins don't get a chance to escape. The more water you use and the longer you take to cook your vegetables, the more vitamins you lose. Steaming, microwaving and stir-frying are by far the best cooking methods to get the best vitamin intake from your veggies.

Table 9-1 shows the characteristics of fat-soluble and water-soluble vitamins.

Table 9-1	Main Differences Between Fat-soluble and Water-soluble Vitamins	
	Fat-soluble Vitamins	*Water-soluble Vitamins*
Risk of deficiency	Very low-fat diets Conditions where fat absorption is poor	Diet lacking variety
Stability in foods	Robust to heat and light	Various; often liable to heat and light
Storage in the body	Can be large and long term	Often small; frequent regular intakes required

Nutritionists use the term *subclinical deficiency* to describe a nutritional deficit not yet far enough advanced to produce obvious symptoms. However, the phrase has become a handy explanation for common but hard-to-pin-down symptoms such as fatigue, irritability, nervousness, emotional depression, allergies and insomnia. And it's a great way to increase the sale of nutritional supplements – call us cynical if you like, we don't care!

Simply put, the RNIs protect you against deficiency. If your odd symptoms linger even after you take reasonable amounts of vitamin supplements, something other than a lack of any one vitamin is probably to blame. Don't wait until your patience or your bank account has been exhausted to find out. Get a second opinion as soon as you can.

Table 9-2 summarises the symptoms of various vitamin deficiencies.

Table 9-2	Vitamin Alert: What Happens When You Don't Get the Vitamins You Need
A Diet Low in This Vitamin	*May Produce These Signs of Deficiency*
Biotin	Loss of appetite; upset stomach; pale, dry, scaly skin; hair loss; emotional depression; skin rashes in infants younger than 6 months
Folate	Anaemia (immature red blood cells)
Niacin	Pellagra (diarrhea; inflamed skin and mucous membranes; mental confusion and/or dementia)

(continued)

Table 9-2 (continued)

A Diet Low in This Vitamin	May Produce These Signs of Deficiency
Riboflavin (vitamin B2)	Inflamed mucous membranes, including cracked lips, sore tongue and mouth, burning eyes; skin rashes; anaemia
Thiamin (vitamin B1)	Poor appetite; unintended weight loss; upset stomach; gastric upset (nausea, vomiting); mental depression; an inability to concentrate
Vitamin A	Poor night vision; dry, rough or cracked skin; dry mucous membranes including the inside of the eye; slow wound healing; nerve damage; reduced ability to taste, hear and smell; inability to perspire; reduced resistance to respiratory infections
Vitamin B6	Anaemia; convulsions similar to epileptic seizures; skin rashes; upset stomach; nerve damage (infants)
Vitamin B12	Pernicious anaemia (destruction of red blood cells, nerve damage, increased risk of stomach cancer attributed to damaged stomach tissue, neurological/psychiatric symptoms attributed to nerve cell damage)
Vitamin C	Scurvy (bleeding gums; tooth loss; nosebleeds; bruising; painful or swollen joints; shortness of breath; increased susceptibility to infection; slow wound healing; muscle pains; skin rashes)
Vitamin D	*In children:* rickets (weak muscles, delayed tooth development and soft bones, all caused by the inability to absorb minerals without vitamin D) *In adults:* osteomalacia (soft, porous bones that fracture easily)
Vitamin E	Inability to absorb fat
Vitamin K	Blood fails to clot

Big trouble: Vitamin megadoses

Can you get too much of a good thing? You bet you can.

Some vitamins are toxic when taken in the very large amounts popularly known as *megadoses*. How much is a megadose? Nobody knows for sure. The general consensus, however, is that a megadose is several times the RDA, but the term is very vague. In 2003 the Food Standards Agency Expert Group on Vitamins and Minerals produced a document called 'Safe Upper Levels for Vitamins and Minerals' stating the maximum dose for each where toxic and undesirable side effects can occur.

The Agency produces a wide range of publications for the public and the food industry. Many of these are available free of charge from Food

Standards Agency Publications. To order copies call 0845 606 0667, or e-mail foodstandards@eclogistics.co.uk.

✔ Megadoses of vitamin A (as retinol) may cause liver damage.

Taken by a pregnant woman, megadoses of vitamin A may damage the fetus.

✔ Megadoses of vitamin D may cause kidney stones and hard lumps of calcium in muscles and organs.

✔ Megadoses of niacin (sometimes used to lower cholesterol levels) can damage liver tissue.

✔ Megadoses of vitamin B6 can cause (temporary) damage to nerves in arms and legs, fingers and toes.

With one exception, the likeliest way to get a megadose of vitamins is to take supplements. It's pretty much impossible for you to cram down enough food to overdose on vitamins D, E, K, C and all the Bs. Did you notice the exception? Right: vitamin A. Liver and fish liver oils are concentrated sources of preformed vitamin A (retinol), the potentially toxic form of vitamin A. Liver contains so much retinol that early 20th-century explorers to the South Pole made themselves extremely ill on seal and whale liver. Cases of vitamin A toxicity also have been reported among children given daily servings of chicken liver (poor mites – daily chicken liver is too much for anyone!).

On the other hand, even very large doses of vitamin E, vitamin K, thiamin (vitamin B1), riboflavin (vitamin B2), folate, vitamin B12, biotin and pantothenic acid appear safe for human beings.

You may not have to go sky-high on vitamin A to run into trouble. In January 2003 new data from a 30-year study at University Hospital in Uppsala (Sweden) suggested that taking a multivitamin with normal amounts of vitamin A may weaken bones and raise the risk of hip fractures by as much as 700 per cent. In this study, a high blood level of retinol from large amounts of vitamin A from food or supplements appears to inhibit special cells that usually make new bone, rev up cells that destroy bone and interfere with vitamin D's ability to help you absorb calcium. Of course, confirming studies are needed, but right now, the RNIs for vitamin A are 600 micrograms per day of vitamin A for women and 700 micrograms per day for men.

Special Circumstances: Taking Extra Vitamins as Needed

The RNIs set by the Department of Health are designed to protect healthy people from deficiencies, but sometimes the circumstances of your life (or your lifestyle) mean that you need something extra. You may need larger amounts of vitamins than the RNIs provide if you fit in with any of the following circumstances.

I'm taking medication

Many valuable medicines interact with vitamins. Some drugs increase or decrease the effectiveness of vitamins; some vitamins increase or decrease the effectiveness of drugs. For example, a woman who's using birth control pills may absorb less than the customary amount of the B vitamins. For more about vitamin and drug interactions, see Chapter 23.

I'm a toddler

The Department of Health recommends that all children up to the age of 5 receive vitamin supplements in the form of vitamin drops containing vitamins A, C and D – three key vitamins that many toddlers don't manage to get enough of. Vitamin drops are currently free to children up to 5 in low-income families. You can also buy special children's vitamins over the counter.

Don't give supplements designed for adults to children because the doses may be too high.

I'm a smoker

You probably have low blood levels of vitamin C. More trouble: chemicals from tobacco smoke create more free radicals in your body. International research organisations, which are often tough on vitamin overdosing, say that regular smokers need to take about 66 per cent more vitamin C – up to 100 milligrams a day – than nonsmokers and should also increase their intake of beta-carotene from vegetable sources like carrots, red peppers, spinach, broccoli and tomatoes. However, if you're a smoker don't use antioxidant supplements to achieve your higher vitamin requirements because this could do more harm than good.

I'm vegan

If you follow a vegan diet (one that excludes all foods from animals, including milk, cheese, eggs and fish oils) you simply can't get enough vitamin D without taking supplements. You may also benefit from extra vitamin C because it increases your ability to absorb iron from plant food. And vitamin B12 enriched grains or supplements are a must to supply the nutrients found only in fish, poultry, milk, cheese and eggs.

I'm a couch potato who plans to start exercising

When you do head for the gym, take it slow, and take an extra dose of vitamin E. A study at the USDA Center for Human Nutrition at Tufts University (Boston) suggests that an 800-milligram vitamin E supplement every day for the first month after you begin exercising minimises muscle damage by preventing reactions with free radicals that cause inflammation. After that, you're on your own: the vitamin doesn't help conditioned athletes whose muscles have adapted to workout stress.

I'm pregnant

Keep in mind that 'eating for two' means that you're the sole source of nutrients for the growing fetus, not that you need to double the amount of food you eat. If you don't get the vitamins you need, neither will your baby.

The RNIs for most of the B vitamins, vitamin K and vitamin E are exactly the same as those for women who aren't pregnant. But when you're pregnant, you need extra:

- **Folate:** Folate protects the child against cleft palate (an opening in the roof of the mouth in which the two sides of the palate didn't join together as the unborn baby was developing) and neural tube (spinal cord) defects. The accepted increase in folate for pregnant women has been 100 milligrams (slightly more than the amount in 4 fluid ounces (113 grams) of orange juice). But new studies show that taking 400 milligrams of folate before becoming pregnant and through the first two months of pregnancy significantly lowers the risk of giving birth to a child with cleft palate. Taking 400 milligrams of folate each day through an entire pregnancy reduces the risk of neural tube defect.

- **Vitamin A:** You need extra vitamin A for your stores and to allow for the baby's growth in the later stages of pregnancy. However, high intakes of vitamin A can be harmful and lead to liver damage. The safe level of supplementation for pregnant women (set by the American College of Obstetrics and Gynaecology in 1993) is 1,500 micrograms per day.

- **Vitamin C:** The level of vitamin C in your blood falls as your vitamin C flows across the placenta to your baby, who may – at some point in the pregnancy – have vitamin C levels as much as 50 per cent higher than yours. So you need an extra 10 milligrams of vitamin C each day for your own health.

- **Vitamin D:** Every smidgen of vitamin D in a newborn's body comes from its mum. If she doesn't have enough D, neither will the baby. Be careful, because too little vitamin D can weaken a developing fetus, but too much can cause birth defects. Check with your doctor who can work out what's right for you according to your weight.

I'm breast-feeding

You need extra vitamin A, vitamin E, thiamin, riboflavin and folate to produce sufficient quantities of nutritious breast milk (about 750 millilitres) each day. You need extra vitamin D, vitamin C and niacin as insurance to replace the vitamins you lose by transferring them to your child in your milk.

I'm approaching menopause or getting on a bit

Older women require extra calcium to stem the natural loss of bone that occurs when you reach menopause and your production of the female hormone oestrogen declines. You may also need extra vitamin D to enable your body to absorb and use the calcium. Gender bias alert! No similar studies are available for older men. But adding vitamin D supplements to calcium supplements increases bone density in older people. The current RNI for vitamin D is set at 10 micrograms per day.

Older people of both genders are also at risk of deficiency of vitamin C, folate and vitamin B12 because they're less likely to eat enough of the foods that are rich sources.

I'm well but worried

A lot of people who are perfectly fit and healthy, eat well and take regular exercise also like to take a vitamin supplement. Nothing's wrong with taking a general multivitamin as an 'insurance policy' for days when you may not eat as well as you'd like. You shouldn't take individual vitamin supplements without consulting your doctor or a dietitian because some vitamins can be toxic at high doses. The idea that the more vitamin you have, the more good it will do isn't accurate: in some cases high doses of vitamins can cause extreme harm.

Check with your doctor before adding vitamin D supplements. In very large amounts, this vitamin can be toxic.

Chapter 10

Mighty Minerals

· ·

In This Chapter

▶ Understanding how your body uses minerals

▶ Getting the minerals you need from food

▶ Finding out what happens when you don't get enough (or too many) minerals

▶ Knowing when you need a little extra

· ·

Minerals are substances that occur naturally in nonliving things such as water, rocks and soil. Minerals also are present in plants and animals, but they're imported: plants get minerals from soil; animals get minerals by eating plants or other animals.

Most minerals have names reflecting the places where they're found or characteristics such as their colour. For example, the word calcium comes from *calx*, the Greek word for lime (chalk); chlorine comes from *chloros*, the Greek word for greenish-yellow. Minerals are *elements*, substances composed of only one kind of atom. Minerals are inorganic; unlike vitamins, they usually don't contain the carbon, hydrogen and oxygen atoms found in all organic compounds.

In this chapter you discover which minerals your body requires to stay in tip-top shape, where you find these minerals in food and precisely how much of each mineral a healthy person needs.

Taking Inventory of the Minerals You Need

Think of your body as a house. Vitamins are like tiny little maids and butlers, scurrying about to turn on the lights and make sure that the windows are closed to keep the heat from escaping. Minerals are more sturdy stuff: the bricks and mortar that strengthen the frame, and the current that keeps the lights on.

An elementary guide to minerals

The early Greeks thought that all material on Earth was constructed of a combination of four basic elements – earth, water, air and fire. Wrong. Centuries later, alchemists looking for the formula for precious metals, such as gold, decided that the essential elements were sulphur, salt and mercury. Wrong again.

In 1669 a group of German chemists isolated phosphorus, the first mineral element to be accurately identified. After that, things moved a bit more swiftly. By the end of the 19th century, scientists knew the names and chemical properties of 82 elements. Today, 109 elements have been identified. Nutritionists consider only 15 minerals to be essential nutrients and 1 semi-essential for human beings.

The classic guide to chemical elements (including minerals) is the periodic table, a chart devised in 1869 by Russian chemist Dmitri Mendeleev (1834–1907). The table was revised by British physicist Henry Moseley (1887–1915), who came up with the concept of *atomic numbers*, numbers based on the number of *protons* (positively charged particles) in an elemental atom.

The periodic table is a clean, crisp way of characterising the elements, and if you know any chemistry, physics or medical students, you can bet your life they've memorised the information it provides. Personally, we'd rather eat overcooked Brussels sprouts.

Nutritionists classify minerals essential for human life as either *major minerals* (including the principal electrolytes, see Chapter 12) and *trace elements*.

For the purposes of this book we define major minerals as those generally required in milligram quantities (sometimes several hundred milligrams) and stored in the body in similar amounts. We define trace elements as those generally required in much smaller quantities, usually microgram amounts, each day.

Some minerals interact with other minerals or with drugs. For example, calcium binds the antibiotic tetracycline into compounds your body can't break apart so that the antibiotic moves out of your digestive tract, unabsorbed and unused. For this reason you should avoid both dairy products and calcium supplements for two hours either side of taking the antibiotic. For more about interactions between minerals and medicines, turn to Chapter 23.

Introducing the major minerals

The following major minerals are essential for human beings:

- ✔ Calcium
- ✔ Phosphorus
- ✔ Magnesium
- ✔ Iron

✔ Zinc

✔ Sulphur

✔ Fluorine (as fluoride)

✔ Sodium

✔ Potassium

✔ Chloride

Note: Sodium, potassium and chloride are also known as the principal electrolytes. We discuss them in Chapter 12.

Calcium

About 1.5 kilograms (3 pounds) of your body weight is calcium, 99 per cent of it packed into your bones and teeth.

The remaining 1 per cent is present in extracellular fluid (the liquid around body cells), where it performs the following duties:

✔ Regulating fluid balance by controlling the flow of water in and out of cells

✔ Making it possible for nerve cells to send messages back and forth from one to another

✔ Keeping muscles contracting normally

✔ Enabling normal blood clotting

An adequate amount of calcium is also important for controlling high blood pressure – and not only for the person who takes the calcium directly. At least one study shows that when a pregnant woman gets a sufficient amount of calcium, her baby's blood pressure stays lower than average for at least the first seven years of life, meaning a lower risk of developing high blood pressure later on.

Getting enough calcium also seems to help with weight loss as well as potentially lowering your risk of cancer of the colon and rectum. The evidence is still inconclusive, but calcium seems to decrease the growth of cells in the colon and reduces the chance that developing cells may become cancerous.

Your best food sources of calcium are milk and dairy products, plus fish such as tinned sardines and salmon where you eat the bones. Calcium is also found in fortified soya milk, bread and cereals, dark green leafy vegetables, pulses and some dried fruits, seeds and nuts. However, the calcium in some plant foods is less well absorbed than that from animal foods and is present in smaller amounts, meaning that you often need to eat quite large amounts to meet requirements (see Table 10-1). Tap water is also a reasonable source of calcium in hard water areas.

Dietary needs for calcium are highest during periods of rapid bone growth such as infancy and teenage years. Needs during pregnancy are generally covered by increased absorption, but requirements do go up by an extra 550 milligrams per day for breastfeeding mothers. The recommended nutritional intakes (RNIs) are 1,000 milligrams per day for 11–18-year-old boys and 800 milligrams per day for girls; and 700 milligrams per day for adults.

In order for your body to absorb calcium, you need an adequate supply of vitamin D from sunlight, your diet or supplements (see Chapter 24).

Table 10-1	Foods Providing Approximately One Third of the Adult RNI for Calcium
Food	*Portion Size*
Milk	1/3 pint (200ml)
Cheese	Small chunk (30g)
Yogurt	1 small carton (150ml)
Cottage cheese	1 large carton (200g)
Canned sardines	1 serving (60g)
White bread	5 large slices
Green leafy vegetables	5 servings (400g)
Baked beans	3 servings (500g)
Mixed nuts	Large portion (300g)
Sesame seeds	3 tablespoons (35g)

Phosphorus

Like calcium, phosphorus is essential for strong bones and teeth: 85 per cent of phosphorus in your body is in your bones. You also need phosphorus to transmit the *genetic code* (genes and chromosomes that carry information about your special characteristics) from one cell to another when cells divide and reproduce. In addition, phosphorus:

- ✔ Helps maintain the pH balance of blood (keeps it from being too acidic or too alkaline)

- ✔ Is vital for metabolising carbohydrates, synthesising proteins and ferrying fats and fatty acids among tissues and organs

- ✔ Is part of *myelin*, the fatty sheath that surrounds and protects each nerve cell

Phosphorus is found in almost everything you eat so deficiency is rare. The best sources are high-protein foods such as meat, fish, poultry, eggs and milk.

These foods provide more than half the phosphorus in a non-vegetarian diet; cereals, nuts, seeds, pulses fruit and vegetables also provide respectable amounts.

Calcium and phosphorus tend to work hand in hand, and an ideal dietary ratio is twice as much calcium as phosphorus. Nutritionists used to believe that a low proportion of calcium to phosphorus in the diet was detrimental in reducing the amount of calcium absorbed and increasing the amount excreted. However, recent studies show this to be untrue unless the phosphorus content of the diet is excessive (on some low-carb/high-protein diets based on meat and fish, for example).

The RNI for adults is 550 milligrams per day, with an extra 440 milligrams for breastfeeding mothers. Like calcium, requirements are higher during growth spurts: 775 milligrams per day for boys aged 11–18 and 625 milligrams per day for girls.

Magnesium

Your body uses magnesium to regulate energy release, nerve cell function and muscle contraction, and to make body tissues, especially bone. The adult human body has about 30 grams of magnesium, and three-quarters of it's in the bones. Magnesium is also part of more than 300 different enzymes that trigger chemical reactions throughout your body.

The main sources of magnesium in the typical UK diet are bread and other cereal products, especially whole grains, followed by beverages such as beer and coffee (no, that's not an excuse for another pint). Milk and meat are also good sources. Many other plant foods are also a rich supply of magnesium, including dark green vegetables (magnesium is part of chlorophyll, the green pigment in plants), seeds, nuts and pulses.

The RNIs for adults over 19 is 300 milligrams per day for men and 270 milligrams per day for women. Breastfeeding women are recommended to take an additional 50 milligrams a day. Deficiency is rare but low intakes can occur in people with restricted diets.

Iron

The adult body contains around 4–5 grams of iron, the majority of which is found as a constituent of haemoglobin and myoglobin, two proteins that transport and transfer oxygen to the cells. You find haemoglobin in red blood cells (it's what makes them red). Myoglobin (*myo* = muscle) is in muscle tissue. Iron is also part of various enzymes and is essential for healthy functioning of the immune system.

The best food sources of iron are offal (liver, heart, kidneys), red meat and meat products, eggs and oily fish. These foods contain haem (*haem* = blood) iron, a form of iron that your body can easily absorb.

Whole grains, fortified cereals, green leafy vegetables, dried fruit, pulses, nuts and seeds contain non-haem iron, which you don't absorb so well. Plants contain fibre and substances called *phytates* and *tannins* that bind this iron into compounds, so that your body has a harder time getting at the iron. However, eating plant foods with meat or with foods that are rich in vitamin C (like green leafy vegetables, tomatoes or fruit juice) can enhance the absorption of iron from plant foods.

The RNIs for iron are 11.3 milligrams per day for boys aged 11–18, 8.7 milligrams for men, 14.8 milligrams a day for women aged 11–50 and 8.7 milligrams for women above 50.

Because the body increases iron absorption at times of biological demand, no specific increments are recommended for pregnancy or breastfeeding. Your doctor or nutritionist may recommend supplements if your iron stores are low or your usual intake is poor.

Zinc

The human body has about 2 grams of zinc stored mainly in muscle and bone and that zinc performs a wide range of roles in the body including normal growth and repair, wound healing and healthy immunity. An adequate supply of zinc is vital for making many enzymes and hormones, including growth hormones, insulin and testosterone, the hormone a man needs to produce plentiful amounts of healthy sperm. Without enough zinc, male fertility falters.

Good sources of zinc are red meat and meat products, milk, eggs, fish and shellfish (yes, the old wives' tale is true: oysters – a rich source of zinc – are useful for men!). Other good sources include wholegrain cereals, green vegetables, pulses and beans, as well as some nuts and seeds. However, like with iron, you absorb the zinc in plants less efficiently than the zinc in foods from animals.

Over the age of 15, the RNI for men is 9.5 milligrams per day and 7 milligrams a day for women. Breastfeeding women should aim for an extra 6 milligrams a day for the first four months and an extra 2.5 milligrams per day until breastfeeding stops.

Sulphur

Although sulphur is often considered an essential nutrient for human beings, nutritional books or charts almost never include the mineral. Why? Because sulphur is an integral part of all proteins as well as being found in fats and many body fluids. Any diet that provides adequate protein also provides adequate sulphur and deficiency is very rare. (Chapter 4 has loads more on proteins.)

Fluorine

Fluorine is found mainly in the form of fluoride in both food and water. Your body stores fluoride in bones and teeth. Researchers often classify it as semi-essential because although a deficiency has never been demonstrated in humans it does have some health benefits. For example, it's clear that incorporating fluorine into dental enamel hardens it, reducing your risk of getting tooth decay. Low water levels are associated with much higher rates of dental caries (decay). In addition, some nutrition researchers suspect (but haven't proved) that you can use some forms of fluoride can to treat osteoporosis.

Small amounts of fluoride occur naturally in all soil, water, plants and animal tissues. You also get a steady supply of fluoride from fluoridated drinking water. About 75 percent of your intake comes from water and only 25 percent from food.

Because it's hard to find proof of a biological need for fluoride, no set RNIs exist, but a safe intake for adults and children over 6 is 0.5 milligrams per kilogram of body weight a day.

Introducing the trace elements

Trace elements are also minerals, but you need them in much, much smaller amounts – microgram (mcg) quantities – in other words, just a trace. Trace elements include:

- Iodine
- Selenium
- Copper
- Manganese
- Chromium
- Molybdenum

Iodine

Iodine is a component of the thyroid hormones thyroxine and triiodothyronine, which help regulate cell activities. These hormones are also essential for protein synthesis, tissue growth (including the formation of a healthy nervous system) and bones.

In the United Kingdom the major source of iodine is milk and milk products. This isn't because it occurs naturally but because cows are often given iodine-rich feedstuffs, and milk is processed and stored in machines and vessels cleaned with iodine-based disinfectants.

The best natural sources of iodine are seafood and plants grown near the ocean, including seaweeds. In some other countries people are most likely to get the iodine they need from iodised salt (plain table salt with iodine added). Products found naturally in some vegetables such as turnips, sweet potatoes, cabbage, millet and corn can inhibit the absorption of iodine. However, in practice this only has a significant effect if your iodine intake is unusually low.

For adults the RNI is set at 140 micrograms per day.

Selenium

Selenium was identified as an essential human nutrient in 1979 when Chinese nutrition researchers discovered that people with low body stores of selenium were at increased risk of *Keshan disease*, a disorder of the heart muscle with symptoms that include rapid heartbeat, enlarged heart and (in severe cases) heart failure, a consequence most common among young children and women of child-bearing age. Selenium deficiency has also been linked to reduced function of the immune system.

In view of the important role of selenium in the antioxidant defence system and immune function of the body, researchers have suggested that selenium may help to protect against heart disease, cancer and other chronic diseases. However, the benefits from increasing intake or long-term supplementation haven't yet been shown in humans.

Fruit, vegetables and cereals grown in selenium-rich soils are themselves rich in this mineral, but in the United Kingdom selenium levels in soil and in many imported crops are relatively low. As a result the best sources of selenium are meat and offal (liver and kidneys), fish, eggs and nuts.

The adult RNIs for selenium are 75 micrograms a day for men and 60 micrograms per day for women, with an extra 15 micrograms per day recommended during breastfeeding.

Copper

Copper is an antioxidant found in enzymes that deactivate free radicals (pieces of molecules that can link up to form compounds that damage body tissues). Copper may therefore be important in helping to protect against cancer and heart disease. Copper also helps to defend the body against infection. Research suggests that copper may even play a role in slowing the ageing process by decreasing the incidence of *protein glycation*, a reaction in which sugar molecules hook up with protein molecules in your bloodstream, twist the protein molecules out of shape and make them unusable. Protein glycation may result in bone loss, high cholesterol and cardiac abnormalities. In people with diabetes, excess protein glycation may also be a factor involved in complications such as loss of vision.

In addition, copper:

✔ Promotes the growth of strong blood vessels and bones

✔ Protects the health of nerve tissue

You can get the copper you need from eating meat, bread and cereals and green vegetables. Other rich sources include liver, shellfish, nuts, tea, coffee and beans, including cocoa beans (bring on the chocolate!). Most other foods, apart from milk and dairy products, provide some copper, so deficiency is rare in the United Kingdom except as a result of severe malnutrition or genetic disorders.

The RNI for adults is 1.2 milligrams a day, with an extra 0.3 milligrams a day needed during breastfeeding.

Manganese

Manganese is an essential constituent of the enzymes that metabolise carbohydrates and synthesise fats (including cholesterol). Most of the manganese in your body is in glands (pituitary, mammary, pancreas); organs (liver, kidneys, intestines); and bones. Manganese is important for a healthy reproductive system. During pregnancy, manganese speeds the proper growth of foetal tissue, particularly bones and cartilage.

You get manganese from whole grains, cereal products, nuts, fruits and vegetables. Tea is also a good source. No RNI exists, but safe intakes are estimated to be greater than 1.4 milligrams per day for adults.

Chromium

Chromium plays several roles in normal metabolism. It's a necessary partner for *glucose tolerance factor* (GTF), a group of chemicals that enhances the action of insulin (an enzyme from the pancreas) to regulate your use of glucose, the end product of metabolism and the basic fuel for every body cell (see Chapter 6). It also seems to be involved in maintaining both normal blood sugar and cholesterol levels.

People with a deficiency of chromium can develop impaired glucose tolerance, which is improved by chromium supplementation. As a result this led people to look at a possible role for chromium in the prevention and treatment of diabetes. Research in the United Kingdom looked at clinical trials of chromium supplementation but found no benefit to glucose tolerance or lipids in adults with type 2 diabetes. Other worldwide studies are inconclusive or also show no benefit, but research has found that excessive doses are detrimental to health. As a result, to date no national diabetes guidelines have adopted chromium supplementation as a valid treatment recommendation. Advice remains to merely ensure adequate dietary intake from a range of sources such as red meat and offal, wholegrain cereals, nuts and seeds, pulses and yeast.

No RNIs are set but safe levels are above 25 micrograms per day for adults.

Molybdenum

Molybdenum (pronounced mo-lib-den-um) is part of several enzymes that metabolise proteins. You get molybdenum from beans and cereals, but local soil levels determine the amount. Cows eat cereal grains, so milk and cheese have some molybdenum. Molybdenum also leeches into drinking water from surrounding soil. Safe intakes for adults are set in the range of 50–400 micrograms per day.

Mineral Overdoses and Underdoses

The RNIs for minerals and trace elements are generous allowances, large enough to prevent deficiency in most healthy people but not so large that they trigger toxic side effects.

Avoiding mineral deficiency

Some minerals, such as phosphorus and magnesium, are so widely available in food that deficiencies are rare to nonexistent. No nutritionist has yet been able to identify a naturally occurring deficiency of sulphur, manganese, chromium or molybdenum in human beings who follow a sensible diet. Most drinking water contains adequate fluoride. Most people get so much copper that deficiency is practically unheard of except in cases of genetic disorders or as a result of severe malnutrition.

But other minerals are more problematic:

✔ **Calcium:** Without enough calcium, a child's bones and teeth won't grow strong and straight, and an adult's bones won't hold on to their minerals. Calcium is a team player. To protect against deficiency, you also need adequate amounts of vitamin D so that you can absorb the calcium you get from food or supplements, and oestrogen, the hormone that helps hold minerals in bone. Foods fortified with vitamin D and people's exposure to sunlight have done much to eliminate rickets (see Chapter 9 on vitamins), and the use of oestrogen replacement therapy – although controversial – does help maintain healthy bones.

✔ **Iron:** Iron deficiency (*anaemia*) is the most common mineral deficiency worldwide. At-risk groups include infants after the liver stores have run out over 6 months of age if they have with delayed or unsuitable weaning or are introduced to cow's milk (a poor source of iron) as their main drink below 1 year of age. They also include menstruating women, pregnant women with low iron stores and vegetarians with few non-haem iron sources in their diets. Lacking sufficient iron, your body cannot make the haemoglobin it requires to carry energy-sustaining oxygen to every tissue. As a result, you're often tired and feel weak. Mild iron deficiency may also inhibit physical development and intellectual performance, especially in children. In one US study, schoolgirls scored higher on verbal, memory and learning test scores when they took supplements providing RNI amounts of iron compared with a placebo.

✔ **Zinc:** Men who don't get enough zinc may be temporarily infertile. Zinc deprivation affects tissues with a rapid turnover and mild deficiency may be relatively common following illness or where intake is low from the diet. Deficiency can make you lose your appetite and your ability to taste food as a result of reduced production of the taste buds. It may also weaken your immune system, increasing your risk of infections. Wounds heal more slowly when you don't get enough zinc.

✔ **Iodine:** Iodine deficiency is now rare in Europe but more common in other parts of the world where it leads to a reduced production of thyroid hormones *and goitre* (a swollen thyroid gland) . The goitre develops as a result of overstimultion of the thyroid gland by the brain which is trying in vain to get it to compensate for falling levels of thyroid hormones in the blood. Other symptoms of iodine deficiency include dry skin , hair loss and tiredness.A severe deficiency in pregnancy leads to an increased rate of stillbirth and in early life may cause a form of mental and physical retardation called *cretinism.*

Table 10-2 lists some of the known results of mineral deficiencies.

Table 10-2	What Happens if You Don't Get the Minerals You Need?
People Who Don't Get Enough of This Mineral	*May Experience These Symptoms*
Calcium	Children: Stunted growth; higher risk of tooth decay; rickets Adults: Low bone density; increased risk of osteoporosis and osteomalacia (adult rickets) usually due to failure to absorb calcium owing to vitamin D deficiency; high blood pressure
Phosphorus*	Deficiency rare; fragile bones and weak muscles; nerve damage.
Magnesium*	Muscle and bone weakness; cramps; high blood pressure; irregular heartbeat; neurological problems including depression and tiredness
Iron	Iron-deficiency anaemia; fatigue caused by lower haemoglobin levels; impaired mental and motor development in children

(continued)

Table 10-2 *(continued)*

People Who Don't Get Enough of This Mineral	May Experience These Symptoms
Zinc	Loss of appetite; reduced ability to taste food; higher susceptibility to infection; retarded growth; slower wound healing; delayed puberty; male infertility
Iodine	Goitre (swollen thyroid gland); mental and physical retardation; slow learning; stillbirth; increased death rate or abnormalities at birth
Selenium*	Muscle pain and weakness (including the heart Keshan disease); reduced immunity
Copper*	Anaemia; impaired immune function; greater susceptibility to infection; weakened bones; changes in hair colour
Manganese*	Dietary deficiency unknown
Fluoride*	Higher risk of dental caries (decay) if low levels in drinking water
Chromium*	No known dietary deficiencies; low intakes may lead to impaired glucose tolerance
Molybdenum*	Dietary deficiency unknown

Dietary deficiency is unlikely to occur in normal healthy people with a varied diet.

Knowing how much is too much

Like some vitamins, some minerals are potentially toxic in large doses:

✔ **Calcium,** though clearly beneficial in amounts higher than the current RNIs, isn't problem free.

- Doses higher than 2–4 grams a day may be linked to kidney damage.
- Megadoses of calcium can bind with iron and zinc, making it harder for your body to absorb these two essential trace elements.

✔ Too much **phosphorus** can lower your body stores of calcium and is harmful in some kidney diseases.

✔ Megadoses of **magnesium** appear safe for healthy people, but if you have kidney disease, the magnesium overload can cause weak muscles, breathing difficulty, irregular heartbeat and/or *cardiac arrest* (your heart stops beating).

✔ **Iron** overload from the diet isn't usually a risk but overdosing on iron supplements can be deadly, especially for young children. The lethal dose for a young child may be as low as 3 grams (3,000 milligrams) elemental iron at one time. This is the amount in 60 tablets with 50 milligrams elemental iron each. For adults, the lethal dose is estimated to be 200 to 250 milligrams elemental iron per kilogram (2.2 pounds) of body weight. That's about 13,600 milligrams for a 70-kilogram person – the amount you'd get in 292 tablets with 50 milligrams elemental iron each. Check with your doctor before downing iron supplements or cereals fortified with 100 per cent of your daily iron requirement.

- Haemochromatosis, a genetic defect affecting one in every 250 people, can lead to _iron overload_, an increased absorption of the mineral linked to arthritis, heart disease and diabetes, and an increased risk of infectious diseases and cancer (viruses and cancer cells thrive in iron-rich blood).

- Moderately high doses of **zinc** (up to 25 milligrams a day) may slow your body's absorption of iron and copper. Doses 27 to 37 times the RNI (11 milligrams for males; 8 milligrams for females) may interfere with your immune function and make you more susceptible to infection, the very thing that normal doses of zinc protect against. More than 2 grams of zinc a day can cause symptoms of zinc poisoning: nausea, vomiting and irritation of the stomach lining.

✔ **Copper** is toxic in high doses – cases have occurred from contamination of water supplies. Dietary excess is unlikely, but take care to avoid large supplemental doses.

✔ **Iodine** overdoses cause exactly the same problem as iodine deficiency – goitre. How can that be? When you consume very large amounts of iodine, the mineral stimulates your thyroid gland, which swells in a furious attempt to step up its production of thyroid hormones. This reaction may occur among people who eat lots of dried seaweed for long periods of time. High intakes are also linked to thyroid cancer and overactive thyroid. Safe upper limits are no more than 1,000 micrograms per day.

✔ Over-supplementation with **selenium** is risky because the difference between safe and toxic doses is relatively small. Safe upper levels are set for adults at no more than 450 micrograms per day. Toxic effects are seen at levels as low as 900 micrograms per day. In China nutritionists linked doses as high as 5 milligrams of selenium a day (90 times the RNI) to thickened but fragile nails, hair loss and perspiration with a garlicky odour. Yuk! In the United States a small group of people who took a supplement that mistakenly contained 27.3 milligrams selenium (436 times the US recommended daily amount) fell victim to selenium intoxication – fatigue, abdominal pain, nausea, diarrhoea and nerve damage. The longer they used the supplements, the worse their symptoms were.

✔ Despite decades of argument, no scientific proof exists that the fluoride in drinking water increases the risk of cancer or bone fracture in human beings. Large doses of fluoride cause fluorosis (mottling and discolouration of the teeth) and possibly some skeletal damage. However, fluoride levels higher than 6 milligrams a day are considered hazardous and those above 500 milligrams a day can be fatal.

Mixing and matching minerals

Taking too much of one mineral can make it difficult for your body to use other minerals.

This list shows which mineral megadoses can affect your ability to absorb and use other minerals and trace elements.

If You Get Too Much of This Mineral	Your Body May Not Be Able to Absorb or Use This One
Calcium	Magnesium, iron, zinc, copper, iodine
Copper	Zinc
Iron	Phosphorus, zinc, copper
Manganese	Iron, iodine
Molybdenum	Zinc, copper
Phosphorus	Calcium, copper
Zinc	Copper, iron, manganese

Munching More Minerals: People Who Need Extra

If your diet provides enough minerals to meet the RNIs, you're in pretty good shape, most of the time.

But a restrictive diet, the circumstances of your reproductive life and just getting older can increase your need for minerals. Here are some scenarios.

You're a strict vegetarian

Vegetarians who avoid fish, meat, poultry and eggs must get their iron from fortified and wholegrain cereals, seeds, nuts, dried fruit, green leafy vegetables, pulses and beans. Because iron in plant foods is bound into compounds that are difficult for the human body to absorb, taking a source of vitamin C such as salad, fruit juice or fruit with your meal can help you absorb the iron.

Vegans – vegetarians who avoid all foods from animals, including dairy products – have a similar problem getting the calcium they need. Calcium is in vegetables, but it, like iron, is bound into hard-to-absorb compounds. So vegans need calcium-rich substitutes. Good food choices are soya milk fortified with calcium, orange juice with added calcium, tofu processed with calcium sulphate, nuts and seeds.

A similar story applies to zinc. Good vegetarian sources include wholegrain cereals, green vegetables, pulses and beans, as well as some nuts and seeds. However, the zinc in plants, like the iron and calcium in plants, is less efficient than the zinc in foods from animals.

Mineral supplements may be useful for some vegetarians and vegans who eat a limited range of foods.

NUTRITION SPEAK

Ferreting out the ferrous stuff in iron supplements

The iron in iron supplements comes in several different forms, each one composed of elemental iron (the kind of iron your body actually uses) coupled with an organic acid that makes the iron easy to absorb.

The iron compounds commonly found in iron supplements are:

✔ Ferrous citrate (iron plus citric acid)

✔ Ferrous fumarate (iron plus fumaric acid)

✔ Ferrous gluconate (iron plus a sugar derivative)

✔ Ferrous lactate (iron plus lactic acid, an acid formed in the fermentation of milk)

✔ Ferrous phosphate (iron plus phosphoric acid)

✔ Ferrous succinate (iron plus succinic acid)

✔ Ferrous sulphate (iron plus a sulphuric acid derivative)

In your stomach, these compounds dissolve at different rates, yielding different amounts of elemental iron. So supplement labels list the compound and the amount of elemental iron it provides, like this:

Ferrous gluconate 300 mg giving iron 34 mg

This tells you that the supplement has 300 milligrams of the iron compound ferrous gluconate, which gives you 34 milligrams of usable elemental iron. If the label just says 'iron', that's shorthand for elemental iron. The elemental iron number is what you look for in judging the iron content of a vitamin/mineral supplement. If you don't like taking tablets, supplemental iron is also available in liquid form.

You're a woman

The average woman of reproductive age loses about 20 milligrams iron a month during menstruation. Women whose periods are very heavy lose more blood and more iron and need more than the RNI. Because getting the iron you need from a diet providing fewer than 2,000 calories a day may be difficult, you may develop a mild iron deficiency. To remedy this, some doctors prescribe a daily iron supplement.

Women who use an intrauterine device (IUD) may also take iron supplements because IUDs irritate the lining of the uterus and cause a small but significant loss of blood and iron.

You're pregnant

The news about pregnancy is that women may not need to take in extra minerals above and beyond the RNIs. Pregnant women still need minerals not only to build foetal tissues, but also new tissues and increased blood volume in their own bodies. However, most of this seems to be provided by increased absorption as a response to increased demand. You may need nutritional supplements if your usual intake or stores are low – speak to your GP or pharmacist for advice.

You're breastfeeding

Nursing mothers need extra iron (met by increased absorption), calcium, phosphorus, magnesium, copper, zinc and selenium to protect their own bodies while producing nutritious breast milk. The same supplements that provide extra nutrients for pregnant women will meet a nursing mother's needs.

You're menopausal

At menopause, your body begins producing less of the bone-protecting hormone oestrogen, and you begin losing bone tissue. You need extra calcium. Severe loss of bone density can lead to osteoporosis and an increased risk of bone fractures, especially if you have a family history of osteoporosis. Women of Caucasian and Asian ancestry are more likely than women of African ancestry to develop osteoporosis. (Men also lose bone tissue as they grow older, but their bones are heavier and denser, and they lose bone tissue less rapidly than women do.)

Chalky-talky: Calcium supplements

Calcium-rich foods give you calcium in a form that your body easily digests and absorbs. The active ingredient of calcium in nearly all supplements in the UK is calcium carbonate, the kind of calcium that occurs naturally in chalk and shells. Calcium carbonate is nearly half calcium, a very high percentage.

You can enhance your absorption of calcium from calcium carbonate by taking the tablets with meals. If you don't like swallowing tablets chewable and liquid forms are available, some of which are pleasantly flavoured to mask any chalky taste.

Because different calcium compounds yield different amounts of elemental calcium, the label should list both the calcium compound and the amount of elemental calcium provided, like this:

Calcium carbonate providing 200 milligrams calcium

Any time you see the word *calcium* alone, it stands for *elemental calcium*. Check that you're getting the right dose for your age and sex (see the earlier section 'Introducing the major minerals') using a manageable minimum number of tablets daily. However, bear in mind that the human body absorbs calcium most efficiently in amounts of 500 milligrams or less. You get more calcium from one 500-milligram calcium tablet twice a day than one 1,000-milligram tablet. If the 1,000-milligram tablets are a better buy, break them in half.

Twenty years ago, nutritionists thought that you couldn't do anything about age-related loss of bone density – that your body stopped absorbing calcium when you passed your mid-20s. Today, everybody knows that increasing your calcium consumption can slow the loss of bone tissue no matter what your age.

Chapter 11

Phabulous Phytochemicals

. . . .

In This Chapter

▶ Explaining what phytochemicals are

▶ Determining why phytochemicals are important

▶ Eyeing accelerating research on phytochemicals

▶ Using phytochemicals every day

. . . .

The latest buzzword in nutrition is *phytochemicals*, a five-syllable mouthful meaning 'chemicals from plants'. Phytochemicals are the substances that produce many of the beneficial effects associated with a diet that includes lots of fruits, vegetables, pulses and grains. This chapter gives you a brief summary of the nature of phytochemicals, lets you know where to find them and explains how they work. For details about how to use phytochemical-rich foods as part of a healthy balanced diet, check out Chapter 21.

Phytochemicals is an area of nutrition where new research is conducted every day. Phytochemicals are likely to play a more significant role in people's diets in the future, but for the moment the only thing we can say with certainty is to eat your veggies and fruit – they really are good for you!

Phytochemicals Are Everywhere

You've been eating phytochemicals all your life without knowing it. The following are all phytochemicals:

- ✔ **Actein,** a hormone-like compound in black cohosh, a native North American herb used by Native Americans and some modern herbalists as a remedy for female troubles such as hot flushes and other signs of menopause

- ✔ **Anthocyanins,** antioxidant pigments such as the colouring agent that makes blueberries blue

- ✔ **Carotenoids,** the pigments that make fruits and vegetables orange, red, yellow, and green

✔ **Daidzein and genistein,** hormone-like compounds in many fruits and vegetables and soya

✔ **Dietary fibre,** the indigestible part of plant foods

✔ **Isothiocyanates,** the smelly sulphur compounds that make you turn up your nose at the aroma of boiling cabbage

These and other phytochemicals such as vitamins (yes, vitamins) perform beneficial housekeeping chores in your body. They:

✔ Keep your cells healthy

✔ Slow down tissue degeneration

✔ Prevent the formation of *carcinogens* (cancer-producing substances)

✔ Reduce cholesterol levels

✔ Protect your heart

✔ Maintain your hormone balance

✔ Keep your bones strong

The potential health-giving benefits of phytochemicals is one reason the Department of Health, the Food Standards Agency and we urge you to have at least five servings of fruits and vegetables and several servings of grains every day.

Did you notice that no minerals appear in the list of phytochemicals? The omission is deliberate. Plants don't manufacture minerals; they absorb them from the soil. Therefore, minerals aren't phytochemicals.

Perusing the Different Kinds of Phytochemicals

The most interesting phytochemicals in plant foods appear to be antioxidants, hormone-like compounds and enzyme-activating sulphur compounds. Each group plays a specific role in maintaining health and reducing your risk of certain illnesses.

Antioxidants

Antioxidants are named for their ability to prevent a chemical reaction called *oxidation*, which enables molecular fragments called *free radicals* to join together, forming potentially carcinogenic compounds in your body.

Antioxidants also slow the normal wear and tear on body cells, so a diet rich in plant foods (fruits, vegetables, grains and beans) is known to reduce the risk of heart disease and may reduce the risk of some kinds of cancer. For example, consuming lots of lycopene (the red carotenoid in tomatoes) has been linked to a lower risk of prostate cancer as long as you mix the tomatoes with an edible oil, which makes the lycopene easy to absorb. Table 11-1 lists the classes of antioxidant chemicals in plant foods.

Table 11-1	Antioxidants in Plants
Chemical Group	*Found In*
Vitamins	
Ascorbic acid (vitamin C)	Citrus fruits, other fruits, vegetables
Tocopherols (vitamin E)	Nuts, seeds, sunflower oil
Carotenoids (pigments)	
Alpha carotene and beta carotene	Yellow and deep-green fruits and vegetables
Lycopene	Tomatoes
Flavonoids (antioxidants found in plants)	
Resveratrol	Grapes, peanuts, tea, wine

Hormone-like compounds

Many plants contain compounds that behave like *oestrogens*, the female sex hormones. Because only animal bodies can produce true hormones, these plant chemicals are called *hormone-like compounds* or *phytoestrogens* (plant oestrogen).

The two kinds of phytoestrogens are:

- Isoflavones, in fruits, vegetables and beans
- Lignans, in grains

The most valuable phytoestrogens appear to be the isoflavones known as *daidzein* and *genistein*, two compounds with a chemical structure similar to *estradiol*, which is the oestrogen produced by mammalian ovaries.

Like natural or synthetic oestrogens, daidzein and genistein can bind to sensitive spots in reproductive tissue (breast, ovary, uterus, prostate) called oestrogen receptors. But phytoestrogens have weaker oestrogenic effects than natural or synthetic oestrogens. It takes about 100,000 molecules of daidzein or genistein to produce the same oestrogenic effect as one molecule of estradiol. Every phytoestrogen molecule that hooks onto an oestrogen receptor displaces a stronger oestrogen molecule. As a result, many researchers believe that consuming isoflavone-rich foods may give women the benefits of oestrogen (lower cholesterol levels, healthy heart, stronger bones and relief from hot flushes) without the possible risk of side effects linked with Hormone Replacement Therapy (HRT).

The best food sources of daidzein and genistein are soya beans and soya products, but these aren't the only foods that contain isoflavones. For example, the main active ingredient in red clover, a traditional herbal remedy for hot flushes, is *formononetin*, an isoflavone that your body converts to daidzein. The active ingredient in black cohosh, another herbal remedy, is also an isoflavone.

Although questions remain about exactly what dose is best, many studies have shown that isoflavones and lignans are safe and useful for human beings and that regularly consuming foods containing isoflavones and lignans, particularly soya, may reduce the risk of heart disease. On the other hand, coumestans, which are up to 100 times as potent as isoflavones, haven't been proven either safe or effective. These common foods provide isoflavones:

• Apples	• Fennel	• Potatoes
• Carrots	• Garlic	• Red clover
• Cherries	• Parsley	• Soya beans
• Dates	• Pomegranates	

These foods provide lignans:

• Cereals

• Linseeds

Sulphur compounds

Pop an apple pie in the oven and soon the kitchen fills with a yummy aroma that makes your mouth water and your digestive juices flow. But boil some cabbage and – yuck! What is that awful smell? It's sulphur, the same chemical that identifies rotten eggs.

Cruciferous vegetables (which get their name from the Latin word for cross, in reference to their X-shaped blossoms) such as broccoli, Brussels sprouts, cauliflower, kale, kohlrabi, mustard seed, onions, radishes, swede, turnips and watercress all contain stinky sulphur compounds and non-nutrient substances that seem to tell your body to rev up its production of enzymes that inactivate and help eliminate carcinogens.

The presence of these smelly sulphurs is thought to explain why people who eat lots of cruciferous veggies generally have a lower risk of cancer. This theory is bolstered by a laboratory experiment in which rats given chemicals known to cause breast tumours were less likely to develop tumours when they were given *sulphorathane*, one of the smelly sulphur compounds in cruciferous veggies.

Other sulphur compounds in cruciferous vegetables are glucobrassicin, gluco-napin, gluconasturtin, neoglucobrassicin and sinigrin. What a mouthful!

Dietary fibre

Dietary fibre is a special bonus you find only in plant foods. You can't get it from meat, fish, poultry, eggs or dairy foods.

Soluble dietary fibre, such as the pectins in apples and the gums in beans, helps mop up unwanted cholesterol and helps lower your risk of heart disease when part of a healthy diet and lifestyle. Insoluble dietary fibre such as the cellulose in fruit skins bulks up stools (faeces) and prevents constipation; moving food more quickly through your gut so food has less time to create substances thought to trigger the growth of cancerous cells. (Turn to Chapter 13 to find out how much dietary fibre you need to get each day and to Chapter 7 to read everything you ever wanted to know about dietary fibre.)

Plant sterols

The effects of sterol-enriched foods have hit both the headlines and the supermarket shelves in recent years. Sterols and stanols are naturally occurring substances found in plants and wood pulp. Evidence suggests that plant sterols and stanols may reduce blood cholesterol levels by reducing cholesterol absorption from the gut. The research so far suggests that regularly consuming sterols and stanols added to margarines, yoghurts, milks and other products may reduce both total cholesterol and LDL (bad) cholesterol. Nutrititionists are still studying the long-term effect of eating large amounts of sterols and stanols over a long period of time.

However effective plant sterols might be in reducing cholesterol, it's still vital that you also stay physically active, be a healthy weight and eat plenty of fruit and vegetables, some oily fish (or a vegetarian supplement) and moderate amounts of saturated fat to protect your heart.

Phenols

You find phenols or phenolic compounds in a wide range of foods from broccoli to almonds. Acid-based and found in plant cells, they're a type of antioxidant that's been linked to protecting the body from many different diseases. As yet the research is too new to make any firm recommendation about consuming foods enriched with phenols, but keep an ear out for them in the future – we think there'll be some interesting findings in the months and years to come. You will have seen from this chapter that there are so many phyto chemicals in all manor of fruits, vegetables and other plant based foods that the bottom line has to be eat more of them and the bigger the variety the greater the health benefit.

Chapter 12

Water Works

In This Chapter

▶ Understanding why you need water

▶ Finding out how best to get water

▶ Recognising and preventing dehydration

▶ Discovering the role of electrolytes

Water is one of the six main nutrients – you could even argue it *is* the main nutrient because the body requires an almost constant supply. Your body is approximately 70 per cent water; an average 70-kilogram male body has about 45 litres of water, making H_2O the most abundant chemical in the body. The exact percentage of water depends on your sex and age. Muscle tissue has more water than fat tissue. Because the average male body has proportionately more muscle than the average female body, he also has more water. For the same reason – more muscle – a young body has more water than an older one. If you really have to, you can live without food for weeks at a time, getting subsistence levels of nutrients by digesting your own muscle and fat. But water's different. Without it, you'll die in a matter of days – even hours in a hot climate. This chapter helps you understand why water is so vital and how you can keep your fluid balance in balance!

Investigating the Many Ways Your Body Uses Water

Water is a solvent. It dissolves other substances and carries nutrients and other material (such as blood cells) around the body, making it possible for every organ to do its job. You need water to:

Fluoridated water: The real tooth fairy

Apart from the common cold, dental decay (*caries*) is the most common human medical problem. Up to 60 per cent of 9-year-olds in the United Kingdom have signs of caries, particularly among low-income groups. You get caries from *Streptococcus mutans*, bacteria that live in dental plaque. The bacteria digest and ferment carbohydrate residue on your teeth – sugars such as sucrose (in plain table sugar), glucose and fructose are the worst offenders – leaving plaque acid that eats away at the mineral surface of the tooth. This eating away is called *decay*. When the decay gets past the enamel to the softer pulp inside the tooth, your tooth hurts.

Brushing and flossing help prevent caries by reducing plaque so that bacteria have less to feed on. Reducing the frequency of sugar-containing foods and drinks can also help. Another way to reduce your susceptibility to caries is to increase your intake of a trace element, fluoride, widely found in the natural environment.

Fluoride is found in food, tea, water, toothpaste and some mouthwashes. It combines with other minerals in the enamel of the teeth to make it less soluble. You get the most benefit if adequate fluoride is available around the time of the formation of the tooth. Benefits continue when the tooth emerges because fluoride inhibits those evil acid-producing bacteria and repairs holes in the enamel.

Despite the still high occurrence of caries in children, the prevalence has decreased by 50 to 70 per cent over the last 30 to 40 years as a direct result of improved fluoridation of drinking water. Currently, around 10 per cent of the population in the United Kingdom has fluoride added to its water supply as a public health measure.

Occasionally, excessive intake of fluoride from a combination of sources can cause a mild white mottling or yellowish brown staining while teeth are developing and accumulating minerals, usually under 8 years old. Because fluoride also concentrates in bones, some people fear that drinking fluoridated water might raise the risk of brittle bones (osteoporosis), cancer and osteoarthritis, but careful studies have failed to find any link.

✔ Digest food, dissolving nutrients so that they can pass through the intestinal cell walls into your bloodstream, and move food along through your intestinal tract

✔ Carry toxins and waste products out of your body via urine from the kidneys

✔ Provide a medium in which chemical reactions such as metabolism occur

✔ Send electrical messages between cells so that your muscles can move, your eyes can see and your brain can think

✔ Act as a lubricant in joints, the eyes and saliva and in mucous membranes

✔ Prevent overheating just like the water in a car radiator – cooling your body when moisture (sweat) evaporates from your skin

Maintaining the Right Amount of Water in Your Body

Of the 45 or so litres of water in an average adult body, about 30 litres or two-thirds is in the *intracellular fluid*, the liquid inside body cells. The remaining 15 litres is in the *extracellular fluid*, which is all the other body liquids, such as:

- ✔ Interstitial fluid (the fluid surrounding the cells)

- ✔ Blood plasma (the clear liquid in blood)

- ✔ Lymph (a clear, slightly yellow fluid collected from body tissues that flows through your lymph nodes and eventually into your blood vessels)

- ✔ Urine in the bladder

A healthy body has just the right amount of fluid inside and outside each cell, a situation that health professionals call *fluid balance*. Maintaining your fluid balance is essential to life. If too little water is inside a cell, the cell shrivels and dies. Too much water, and the cell bursts.

A balancing act: The role of electrolytes

Your body regulates its fluid balance through the action of substances called *electrolytes*, essential minerals that dissolve in water into electrically charged particles called *ions* and that are found in every body fluid.

Many minerals, including calcium, phosphorus and magnesium, form compounds that dissolve into charged particles. But nutritionists generally use the term *electrolyte* to describe sodium, potassium and chlorine (as chloride). The most familiar source of electrolytes is the one you find on every dinner table – sodium chloride, plain old table salt. (In water, sodium chloride molecules dissolve into two ions: one sodium ion and one chloride ion.)

Under normal circumstances, the fluid inside your cells has more potassium than sodium and chloride. The fluid outside is just the opposite: more sodium and chloride than potassium, rather like seawater. The cell wall is a *semipermeable membrane*; some things pass through, but others don't. Water molecules and electrolytes flow through freely, but larger molecules such as proteins don't. Changes in fluid balance alter the concentration of these electrolytes, either diluting them or concentrating them, which in turn controls the body's response to loss of water.

The body uses this changing concentration of electrolytes to draw water between the cells and maintain a balance between the extra and intra cellular areas. When your body loses too much fluid, the increased concentration of electrolytes in the blood, especially sodium, stimulates your thirst centre, causing you to drink more to replace the lost fluid.

Concentration of the plasma sodium also results in reduced urination, a protective mechanism triggered by *antidiuretic hormone* (ADH), a hormone secreted by the hypothalamus, a gland at the base of your brain, which reduces the release of urine from the kidneys. If you drink too much the concentration of sodium goes down and the reverse occurs – you stop feeling thirsty and start producing more urine.

Dehydrating without enough water and electrolytes

Drink more water than you need, and the healthy body simply shrugs its shoulders, urinates more copiously and readjusts the water levels. A healthy person on a normal diet would find it hard to drink himself or herself to death on water. However, some people with heart or kidney problems may be at risk of fluid overload and need to follow medically supervised fluid restrictions. Some endurance athletes and those on severe 'detoxing' regimes are also at risk from excessive consumption of plain water, leading to very low blood levels of sodium, which can lead to death.

If you don't get enough water, your body lets you know pretty quickly. Rapid fluid depletion can occur in children with diarrhoea and can be fatal.

Electrolytes and minerals

In addition to keeping fluid levels balanced, sodium, potassium and chloride (the form of chlorine found in food) ions create electrical impulses that enable cells to send messages back and forth between themselves, so you can think, see, move and perform all the bioelectrical functions that you take for granted.

Sodium, potassium and chloride are also major minerals (see Chapter 10) and essential nutrients. They're useful in these bodily processes:

- ✔ Sodium helps with nerve impulse transmission and keeps your blood from becoming too acid or too alkaline.

- ✔ Potassium is used in muscle contraction and nerve impulse transmission.

- ✔ Chloride is a constituent of hydrochloric acid, which breaks down food in your stomach.

Chronic dehydration can be a problem for people if they have a low fluid intake. Children (who have a large surface area through which they may lose water), breastfeeding mothers, active adults, older adults with a reduced sense of thirst and those unwell with fever, vomiting or diarrhoea are at increased risk. People on diuretic drugs (medication that increases the flow of urine) or those with swallowing difficulties (dysphagia) also need to be careful.

Recent studies show that as many as one in four people don't drink enough. The thirst mechanism isn't really very sensitive in humans – by the time you feel thirsty, you've already lost about 1 to 2 per cent of your body weight as water and you may already be really quite dehydrated because it can take time for the receptors in the brain to detect an increase in concentration of the blood.

Signs of dehydration include:

- ✔ Concentrated dark-yellowish-brown coloured urine (urine should be pale yellow straw coloured)
- ✔ Constipation
- ✔ Delayed reaction time
- ✔ Exacerbated existing mental confusion
- ✔ Fatigue and irritability
- ✔ Headache
- ✔ Impaired coordination and accuracy
- ✔ Impaired short term memory
- ✔ Loss of appetite
- ✔ Nausea or dizziness
- ✔ Poor concentration or decision-making ability
- ✔ Reduced muscular strength and physical endurance

Chronic dehydration can lead to increased risk of kidney stones or urinary tract infections and may increase the risk of bowel and bladder cancer and stroke. Recent research has also shown chronic dehydration may delay healing of wounds such as bed sores and increase wound infection, especially in older adults. Severe dehydration can lead to circulatory collapse, and even death.

Getting the Water You Need

Unlike other nutrients, you don't store water, so you need to take in a new supply every day, enough to replace what you lose. Total daily water turn-over for a reasonably sedentary individual in a temperate climate is about

2,800 millilitres a day from food and liquid. The average person in Britain uses 150 litres of water each day but only actually drinks 1 litre. You lose about 1,400 millilitres or nearly 50 per cent of water in your urine. You lose another 150 millilitres in your faeces, around 850 millilitres as sweat and 400 millilitres in breath from your lungs.

You don't need to replace all this lost water with drinks. You create about 10 per cent of the water you need (around 300 millilitres) when you digest and metabolise food. The end products of digestion and metabolism are carbon dioxide (a waste product that you breathe out of your body) and water (composed of hydrogen from food and oxygen from the air that you breathe). Around a litre of water comes from your daily food intake. Fruit vegetables, soups and stews are full of it. Lettuce, for example, is 90 per cent water. Even foods such as cooked potatoes and pasta are around 70 per cent water. The rest of your fluid intake comes directly from drinks – around 1,500 millilitres a day. This translates to about six to eight large glasses or cups a day. In a temperate climate, or if you're very active, you need more: you can lose over 2 litres of sweat per hour. Yuck!

Here are some sources of fluid:

How does water know where to go?

Osmosis is the principle that governs how water flows through a semipermeable membrane such as the one surrounding a body cell.

Here's the principle: water flows through a semipermeable membrane from the side where the liquid solution is least dense to the side where it's denser. In other words, the water, acting as if it has a mind of its own, tries to equalise the densities of the liquids on both sides of the membrane.

How does the water know which side is more dense? That's easy: wherever the sodium content is higher. When more sodium is inside the cell, more water flows in to dilute it. When more sodium is in the fluid outside the cell, water flows out from the cell to dilute the liquid on the outside.

Osmosis explains why drinking salty seawater doesn't hydrate your body. When you drink seawater, liquid flows out of your cells to dilute the salty solution in your intestinal tract. The more you drink, the more water you lose. When you drink seawater, you're literally drinking yourself to death.

The same thing happens to a much lesser degree when you eat salted crisps or nuts. The salt in your mouth makes your saliva saltier. This draws liquid out of the cells in your cheeks and tongue, which feel uncomfortably dry. You need a drink – maybe that's why some pubs give out free nuts!

✔ **Simple water:** Tap water is the most easily obtained and cheapest source of fluid in countries such as the UK where its safety is assured. If you dislike tap water, a water filter is a good idea or a slice of lemon improves flavour. Bottled water is a good alternative when you're unsure of the safety of the tap water. However, bottled water can cost up to 100 times more than the equivalent volume of tap water, it has numerous associated environmental costs and it offers no nutritional advantages over tap water (see the later sidebar 'Water, water, everywhere' for more).

✔ **Caffeinated drinks and alcohol:** Here's an interesting fact: not all liquids are equally rehydrating. The caffeine in strong coffee and energy drinks, and the alcohol in beer, wine and spirits, acts as a diuretic, making you urinate more copiously, so these drinks are less useful for rehydration. But don't ditch your espresso machine: recently careful fluid balance studies show that standard servings o coffee have very little diuretic effect in regular consumers who are used to the drinks. Normal black tea as well as herbal, green and fruit teas are all popular and are an excellent way of boosting fluid intake.

✔ **Soft drinks:** Diluted squash and fizzy drinks provide fluid, but along with it comes sugar. Bacteria in the mouth can break down the sugar to form plaque acid, which in turn can cause dental caries (see the earlier sidebar 'Fluoridated water: The real tooth fairy'). The more frequently you bathe teeth in sugary solutions, the more likely decay is to occur. Sugar in drinks can also add to calorie intake. Recently the USA has produced Healthy Beverage Guidelines that recommend less than 10 per cent of your energy should come from drinks high in sugar with few other nutrients. However, currently no such recommendations exist in the UK. Sugar free or diet fizzy drinks are a low a calorie alternative, although they're still acidic and as such have the potential to wear away tooth enamel. Drinking through a straw can help reduce the contact time between a sugary drink and the teeth.

✔ **Fruit juices and smoothies:** Unsweetened 100 per cent fruit or vegetable juices and smoothies count towards fluid intake and provide one of your five-a-day fruit or veg portions. However, these drinks also contain calories from sugar and are acidic. Diluting juice with water is helpful, especially for children.

✔ **Milk :**Milk counts towards your daily fluid intake whether drunk cold on its own or in smoothies or milk shakes flavoured with fruit or as a hot bedtime drink. It's a great source of nutrients but if you are watching your fat intake go for the lower fat versions such as skimmed, semi skimmed and the newer 1% fat milks.

Here are some tips for encouraging fluid intake:

- Choose a variety of drinks, introducing some enjoyable new tastes.
- Try adding in one or two extra drinks each day, and then add another couple after that. After a while you'll develop a regular healthy hydration habit.
- Take regular drinks with meals and make time for drink breaks, especially at work.
- Remind older adults and young children in particular to drink regularly before they're thirsty, especially in hot weather.
- Make drinking fun for children – suggest they try coloured straws and attractive water bottles.

Taking in Extra Water and Electrolytes as Needed

Most people regularly consume much more sodium than they need. Sodium occurs naturally in many foods and is added in a variety of forms including sodium chloride and monosodium glutamate to foods during processing. In fact, high salt intakes may be responsible for high blood pressure, especially the increase in blood pressure associated with getting older. Studies suggest that many people can lower their blood pressure if they reduce their sodium intake. (For more about high blood pressure, check out *High Blood Pressure For Dummies* by Dr Alan Rubin, published by Wiley).

Potassium and chloride are found in so many foods that dietary deficiency is virtually non-existent. However, research has shown that increasing dietary potassium can help to reduce high blood pressure. Particularly useful sources of potassium include fruit, vegetables and potatoes.

The daily reference nutrient intakes (RNIs) for adults for sodium, potassium and chloride are:

- **Sodium:** 1,600 milligrams
- **Potassium:** 3,500 milligrams
- **Chloride:** 2,500 milligrams

Most people get much more of these minerals as a matter of course, but sometimes you actually need extra water and electrolytes.

You've got an upset stomach

Repeated vomiting or diarrhoea drains your body of water and electrolytes. The normal faecal loss of water of 150 millilitres can increase to 1–2 litres with severe diarrhoea. Similarly, you also need extra water to replace the liquid lost in sweat when you have a temperature. When you need fluid in a hurry, plain water won't replace fluid as quickly as a rehydration solution with added electrolytes. Check with your pharmacist for a drink to hydrate your body without upsetting your stomach.

You're exercising or working hard in a hot environment

When you're warm, you sweat. The moisture evaporates and cools your skin so that blood circulating up from the centre of your body to the surface is cooled. The cooled blood returns to the centre of your body, lowering the temperature (your _core temperature_) there too. If you don't replace the water lost in sweat things can get tricky. At first it's just your concentration and ability to keep going that's affected, but ultimately dehydration can lead to heat stroke and circulatory collapse.

Always make sure you're well hydrated when you start exercising. For most recreational or competitive exercise, aim to drink up to 600–800 millilitres of water per hour at regular intervals to top up – don't wait until you feel thirsty. Rehydrate afterwards as soon as you can – and certainly before you head for the bar! Plain water is fine in most situations, but for exercise lasting more than an hour some isotonic sports drinks containing sugars and electrolytes may increase your fluid absorption and offer added carbohydrate for energy. However, sports drinks can be expensive to buy in the amounts you need for rehydration if you're exercising a lot. Homemade versions are cheaper and often even nicer. Try this simple recipe:

1. Mix 250 milliletres of pure unsweetened fruit juice (any flavour) with 250 milliletres of water.

2. Add a pinch of salt (about ⅛ teaspoon).

3. Stir or shake well. Chill slightly.world.

Any flavour of full sugar fruit squash can also be used with 400 milliletres of water.

Water, water, everywhere

Water is the only substance on Earth that can exist as a liquid (water) and a solid (ice or snow) and a gas (steam). Water can be hard or soft depending on the mineral content.

✔ **Hard water** is water that rises to the Earth's surface from underground springs. It has lots of minerals that it picks up as it moves up through the ground. Much of the water in south-east of England is hard.

Hard water may contain as many as 100 particles of calcium, magnesium, iron and sodium for every 1 million parts of water (shorthand: 100 ppm) and can leave a deposit on the bath or a scale in your kettle and washing machine.

✔ **Soft water** is usually surface water collected from streams and hills, or rainwater that falls directly into reservoirs. Wales, Scotland, central and south-west England have soft water. Soft water has fewer minerals.

If you aren't sure about whether your drinking water is hard or soft, consider the source. In areas of the United Kingdom where the major part of the water supply comes from reservoirs, the water that flows from your tap is likely to be soft. In the other parts of the UK, the water is likely to be hard.

Tap water in the United Kingdom meets rigorous chemical and microbiological safety standards set and monitored by the Drinking Water Inspectorate, making it entirely suitable as your main source of fluid. It also has a fraction of the environmental costs associated with the production, transport and packaging of bottled water. However, some people either don't like the taste of their tap water, or they may wish to avoid added chemicals such as fluoride, chlorine used to kill microorganisms or traces of chemicals used in farming. As a result, the last decade has seen a massive explosion in the use of mains and jug water filters and bottled water. In the United Kingdom people spend well over £2 billion per year on bottled water. Who could have predicted that one day entire aisles of supermarkets would be devoted just to bottled water? Here's what you can get:

✔ **Bottled water** is usually tap water that's purified to remove all chemicals by methods such as distillation, ionisation or reverse osmosis. If carbon dioxide is added the water is known as sparkling water.

✔ **Natural mineral water** must come from a natural underground source. This water contains minerals leached from the rocks as it passes through. Generally, the levels of minerals such as calcium are quite low – often not much higher than tap water and only about 10 per cent of that found in milk. Check the label if you're watching your salt intake because some brand-name mineral water can be higher in sodium. For this reason don't use natural mineral water when making up infant formula.

✔ **Spring water** is water from a natural spring source. It can be fizzy or still and often has fewer mineral particles and a cleaner taste than mineral water. Some brands can still be major sources of sodium – always check the label.

If we really can't persuade you away from bottled water then why not see whether you can find one of the not-for-profit bottled waters that use recyclable plastic bottles and donate money to clean water projects around the world.

Part III
Healthy Eating

'Sacrifices are just not the same since the
tribe went vegetarian.'

In this part . . .

You'll find out how to build a healthy diet right here. We explore the key guidelines for a wholesome diet and guide you through easy and tasty ways to put these wise food choices into practice. We include strategies for selecting food that optimises your nutrition, whatever stage of life you're at.

Chapter 13

What Is a Healthy Diet?

In This Chapter

▶ Introducing dietary guidelines

▶ Establishing a healthy diet

▶ Striving for fitness

*T*he British Heart Foundation says to limit your consumption of fats and cholesterol. The World Cancer Research Fund says to eat more fibre, fruit and vegetables. The Food Standards Agency says to watch out for fats, sugar and salt. Diabetes UK says to eat regular meals so that your blood sugar stays even. The Food Police say if it tastes good, forget it!

The Ministry of Agriculture, Fisheries and Food (now the Food Standards Agency) and the Health Education Authority incorporated virtually all but the 'tastes good, forget it' rule into the 1997 *Guidelines for a Healthy Diet*, and even added some more of their own. As a result, the guidelines are to punitive food rules what Häagen-Dazs low-fat sorbet is to ordinary ice cream: a delicious, reasonable and totally guilt-free alternative. We explore those guidelines in this chapter.

Introducing the Guidelines for a Healthy Diet

The *Guidelines for a Healthy Diet* are a collection of sensible suggestions first published by the Ministry of Agriculture, Fisheries and Food and the Health Education Authority in 1991, and revised in 1997. The guidelines describe food and lifestyle choices that promote good health, provide the energy for an active life and may reduce the risk or severity of chronic illnesses, such as diabetes and heart disease.

The best thing about these guidelines is that they seem to have been written by real people who actually like food. You can see this good attitude to food from the word go in the very first paragraph, which begins: 'Eating should be a pleasant aspect of life.' Hallelujah!

The guidelines work best in conjunction with the Eatwell Plate, which groups foods into categories and suggests the proportions of each you need to consume every day. You can read more about the Eatwell Plate in Chapter 14. Right now, however, the job is spelling out the guidelines themselves.

The *Guidelines for a Healthy Diet* deliver eight basic health messages:

- ✔ Enjoy your food.
- ✔ Eat a variety of different foods.
- ✔ Eat the right amount to be a healthy weight.
- ✔ Eat plenty of foods rich in starch and fibre.
- ✔ Eat plenty of fruit and vegetables.
- ✔ Don't eat too many foods that contain a lot of fat.
- ✔ Don't have sugary foods and drinks too often.
- ✔ If you drink alcohol, drink sensibly.

And the Scientific Advisory Committee on Nutrition (2003) report suggests we add another guideline:

- ✔ Choose and prepare foods with less salt.

Throughout this chapter, we expand on what each of these suggestions means.

Stepping Out Towards a Healthier Lifestyle

Healthy eating doesn't mean that any foods are completely banned or that others are obligatory. In order to maximise health and minimise the risk of disease you need to achieve nutritional balance. That means you don't need to restrict foods, you don't need to live in 'nutritional purgatory' and sometimes a little of what you fancy does you good!

Eat a variety of foods

The greater the variety of foods you eat, the more likely that your diet contains all the essential nutrients (including vitamins and minerals) that you need to be healthy. Mother Nature is pretty good at giving you everything you need in the foods you eat. The Eatwell Plate (Chapter 14) shows the perfect model for what a healthy balanced diet should look like.

Eat the right amount for a healthy weight

Being overweight leads to many health problems. It places greater stress on the bones and joints, and raises blood pressure and cholesterol. It worsens breathing difficulties and increases the risk of developing diabetes, heart disease, stroke and some forms of cancer. Eating the right diet and being physically active are essential to maintain a healthy weight.

So how do you go about managing your weight?

- ✔ **Evaluate your body weight.** The best test of whether or not you're overweight is the Body Mass Index (BMI), a measure of body fat versus lean tissue or muscle that predicts the risk your weight poses to your health. You can read about the BMI in more detail in Chapter 6.

- ✔ **Manage your weight.** Weight management is about a lifelong lifestyle approach to food and activity. Quick fixes that sound too good to be true usually are!

- ✔ **If you need to lose weight do it gradually.** Forget the 'lose 2 stone in a month' headlines. Depending on how much weight you have to lose, anything between ½ and 2 pounds a week is a safe, maintainable weight loss. If you're overweight, losing just 10 per cent of your body weight brings significant health benefits, so it's not all about getting down to the weight you were at 16!

- ✔ **Encourage healthy weight in children.** It's sad but true that overweight kids usually become overweight adults. Helping children keep to a healthy weight from the start reduces the likelihood of them having weight problems later.

Eat plenty of food rich in starch and fibre

Contrary to popular belief, foods like bread and potatoes aren't necessarily fattening and provide essential nutrients such as vitamins and fibre. Other starchy, fibre-rich foods like cereals, pasta and rice are filling and relatively cheap. These types of foods should be a major part of each meal and your diet as a whole. (Skip to Chapter 7 for heaps more on carbohydrates.)

Let the Eatwell Plate guide your food choices

The Eatwell Plate is a guide that shows you exactly how to fill your plate. You can use it to help you plan a daily or weekly menu. Chapter 14 is devoted almost entirely to the Eatwell Plate and its virtues as a meal planner. So let's not spend time on it here, except to say that the Eatwell Plate is bang on target for promoting healthy eating.

Eat plenty of fruit and vegetables

Fruits and vegetables are special because they:

- ✔ Add plenty of bulk but few calories to your diet, so you feel full without adding weight

- ✔ Are usually low in fat and have no cholesterol, which means that they reduce your risk of heart disease

- ✔ Are high in fibre, which reduces the risk of heart disease; prevents constipation; reduces the risk of developing haemorrhoids (or at least makes existing ones less painful); moves food quickly through your digestive tract, thus reducing the risk of diverticular disease (inflammation caused by food getting caught in the folds of your intestines and causing tiny pouches of the weakened gut wall); and lowers your risk of cancer of the mouth, throat (oesophagus and larynx) and stomach

- ✔ Are rich in beneficial substances called phytochemicals and antioxidants, also believed to reduce your risk of heart disease and cancer (for more about these wonder workers, see Chapter 11)

For all these reasons, the British Dietetic Association, the British Heart Foundation and just about every other agency with an interest in health recommends that you eat at least five portions of different-coloured fruit and vegetables every day.

People are often confused about what constitutes a portion, so here's a guide to help:

- ✔ A cereal bowl full of mixed, undressed salad

- ✔ A handful of berries such as strawberries or grapes

- ✔ A small glass (150 millilitres or 5 fluid ounces) of unsweetened fruit or vegetable juice – sorry, you can only count fruit juice once a day!

- ✔ Half an avocado

- ✔ One big slice of large fruit such as a large slice of melon or pineapple

- ✔ One medium fruit such as a banana or an apple

- ✔ Three heaped tablespoons of beans (pulses also only count once a day)

- ✔ Two small pieces of fruit such as two plums or two apricots

- ✔ Two tablespoons of dried, cooked or tinned fruit in natural juice

- ✔ Two tablespoons of raw or cooked vegetables

Walking Away from Unhealthy Food

Nutritionists don't like talking about 'good' or 'bad' foods, but you can benefit from eating less of certain foods. Although a little of what you fancy really can do you good, too much of a good thing upsets the balance.

Don't eat too many fatty foods

Eating too much fat tends to raise blood cholesterol levels and increases the risk of obesity and heart disease. Most people would benefit from eating less fat in general and in particular less saturated fat. With fats like butter, margarine and cooking oils, and fried food, full-fat dairy products and fatty meat, you can usually see the fat so it's easy to cut down. A lot of foods such as pastry, pies, biscuits, cakes and savoury snacks contain a lot of hidden fat that you can't see so easily. Check the label for the fat content in foods like these (we talk about food labels in detail in Chapter 14).

No foods are inherently good or bad, and fats are a good example. Fat is an essential nutrient but certain fats, especially the saturated fats mainly found in animal foods, increase blood cholesterol levels. Other types such as monoun-saturated fat, polyunsaturated fat and omega 3 and 6 fatty acids are positively healthy. You've guessed it: it's all a question of balance.

Overall, the *Guidelines for a Healthy Diet* suggest that adults should derive no more than 30 to 35 per cent of calories from fat, with no more than 10 percent of calories from saturated fat. For more information, head to Chapter 5 and read all about fats.

Counting the fat calories

To calculate the percentage of fat in any food, you need to know two numbers: the total calories in the serving and the number of grams of fat. Take, for example, one wedge of Camembert cheese. The label says that the cheese wedge has 115 calories and 9 grams of fat. One gram of fat has 9 calories. Use the following equation to find out the percentage of calories that come from fat:

1. **Multiply the number of grams of fat by 9 (the number of calories in one gram of fat).**

   For the Camembert cheese example, 9 grams multiplied by 9 calories per gram gives you the number of calories from fat, or 81.

2. **Divide the result from Step 1 by the total number of calories. The result is the percentage of calories from fat.**

   Continuing the cheese example, divide 81 (the number of calories from fat) by 115 (the total number of calories in the wedge) and multiply by 100. The result: 70 per cent of the calories come from fat in one wedge of Camembert cheese.

The good news is that most packaged food sold in the UK carries nutritional information, including the fat content. Milk, peas, soup, chocolate cake – you name it, and you can find the total and saturated fat content per serving right there on the food label, which means you can also figure out the percentage of calories from fat.

Don't have sugary foods and drinks too often

Sweet foods and drinks are fine as occasional treats, but too many, too often are likely to put your diet in an energy or calorie excess, which means you're likely to put on weight. If you're substituting fruits, vegetables and starchy high-fibre foods with lots of sweet foods and drinks, you're missing out on essential vitamins and minerals – too much sugar, too often is a double whammy for unbalancing your diet.

Frequent intakes of sugar-rich foods and drinks can also cause tooth decay. Brush teeth twice a day with a fluoride toothpaste and floss regularly. Chewing sugar-free gum can also stimulate the saliva flow to protect the teeth.

Question: which is worse for your teeth, a piece of chocolate or a couple of raisins? Yes, this is a trick question. Yes, the answer is raisins.

The explanation's simple. Both foods have lots of sugar, but the chocolate is less detrimental to your teeth because it dissolves quickly and is washed out of your mouth (and off your teeth) by your saliva. Raisins, on the other hand, are sticky. Unless you brush and floss thoroughly they cling to your teeth, providing a longer-lasting banquet for those pesky tooth-decay bacteria.

Although sugars occur naturally in fruits and vegetables, the safe bet is that most of the sugar in your diet comes as 'added sugar' in processed foods. These added sugars may be listed on food labels in any of the forms in the list that follows. You can assume the food is high in sugar whenever the sugar word is one of the first ingredients listed on the food product's ingredient list.

- Brown sugar
- Corn sweetener
- Corn syrup
- Fructose
- Fruit juice concentrate
- Glucose (dextrose)
- High-fructose corn syrup
- Honey
- Invert sugar
  (50:50 fructose–glucose)
- Lactose
- Maltose
- Molasses
- Raw sugar
- Sugar (sucrose)
- Syrup

If you drink alcohol, drink sensibly

Modest amounts of alcohol aren't harmful to most people and may even have health advantages in some circumstances. Men and postmenopausal women seem to get some reduced risk of heart disease from drinking in moderation. But what's moderation, anyway? Nutritionists define moderate as the amount of alcohol your body can metabolise without increasing your risk of serious illness such as cancer or liver damage.

The government defines sensible drinking limits as three to four units per day for men or two to three units per day for women. However, the government also says that consistently drinking four units a day (three for women) isn't recommended because of the progressive health risk it carries. For more information on the health risks of drinking too much alcohol see Chapter 8.

One unit is:

✔ Half a pint of ordinary strength beer lager or cider

✔ One small glass of wine (100 millilitres)

✔ One single measure of spirits (25 millilitres)

✔ One small glass of sherry (50 millilitres)

The safe levels for women are lower than for men because women are on average smaller than men and their bodies contain proportionately less water and more fat. They can tolerate less alcohol before damage occurs to their organs and can feel its effects faster and more intensely. The old wives' tale that he can drink her under the table is no myth. It's physiology.

Some people shouldn't drink at all, not even in moderation, including people who suffer from alcoholism, people who plan to drive a car or take part in other activities that require attention to detail or physical skill, and people using certain types of medication (prescription drugs or over-the-counter products). Chapter 8 has all the facts on alcohol.

Choose and prepare foods with less salt

Sodium is a mineral that helps regulate your body's fluid balance, the flow of water into and out of every cell (see Chapter 10). This balance keeps just enough water inside the cell so that it can perform its daily jobs, but not so much that the cell explodes.

Most of people get far more sodium than they need, mainly from salt or sodium chloride in their diets. As a result high blood pressure is very common, and this in turn can increase the risk of heart disease and stroke. Salt or sodium makes your blood hold on to more water, which creates more pressure. If you already have high blood pressure, you can help reduce it by lowering the amount of salt you eat. If your blood pressure is normal you can help keep it that way by watching your salt intake.

For a few people reducing salt intake has another, unadvertised benefit. It may lower weight a bit. Why? Because sodium is hydrophilic (*hydro* = water; *philic* = loving). Sodium attracts and holds water. Eating less salt means that some people retain less water and feel less bloated.

Don't reduce salt intake drastically without first checking with your doctor. Remember, some sodium is essential, and the guidelines advocate moderate use, not no use at all.

The obvious question: what's moderate use? Current intakes are in the region of 9 grams of salt (3,600 milligrams of sodium) per day – that's about 1½ teaspoons. The *Guidelines for a Healthy Diet* recommend you aim for no more than 6 grams of salt (2,400 milligrams of sodium) or 1 teaspoon per day. (One gram of salt = 400 milligrams of sodium. To convert sodium to salt, multiply the sodium content by 2.5.)

Like sugar, sodium occurs naturally in foods. However, only about 15 per cent of sodium intake comes via this route. Another 15 per cent is added at the table or in cooking. A whopping 70 per cent comes from manufactured or processed foods. For example, a portion of frozen peas cooked in unsalted water has only a trace of sodium, but a portion of tinned peas has over 200 milligrams of sodium.

Look out for tinned and processed vegetables in lower-salt versions. The difference is notable: a tin of reduced-salt baked beans has approximately half the salt content of regular baked beans.

You also get salt from fast foods, sauces and pickles. Not all the sodium you eat is sodium chloride. Sodium compounds are also used as preservatives, thickeners and flavour enhancers, and these all add to the load.

To find out more about salt check out www.salt.gov.uk.

Table 13-1 lists several different kinds of sodium compounds in food.

Table 13-1	The Sodium Compounds Found in Food
Compound	*What It Is or Does*
Monosodium glutamate (MSG)	Flavour enhancer
Sodium benzoate	Food preservative
Sodium caseinate	Thickens foods
Sodium chloride	Table salt (flavouring agent)
Sodium citrate	Keeps drinks fizzy
Sodium hydroxide	Makes it easy to peel the skin off tomatoes and fruits before they're tinned
Sodium nitrate/nitrite	Keeps food preserved and gives them their distinctive red colour (e.g. cured meats)
Sodium saccharine	Artificial sweetener

Source: 'The Sodium Content of Your Food', Home and Garden Bulletin, No. 233 (Washington, D.C.: U.S. Department of Agriculture, August 1980); Ruth Winter, A Consumer's Dictionary of Food Additives (New York: Crown, 1978)

Relaxing Once in a While

Life isn't a test. You won't fail if you don't manage to follow the *Guidelines for a Healthy Diet* every single day of your life. Nobody's perfect, and the guidelines are meant to be broken once in a while. For example, ideally you should keep your daily fat intake to around 35 per cent of your total calories. But you can bet that you'll exceed that amount this Saturday as you stroll up to the buffet table at your best friend's wedding and see:

✔ Camembert cheese (70 per cent of the calories from fat)

✔ Sirloin steak (56 per cent of the calories from fat) and salad with Thousand Island dressing (90 per cent of the calories from fat)

✔ Hot chocolate fudge cake and cream (we can't count that high)

Is this a crisis? Should you stay at home? Must you keep your mouth shut tight all night? Are you joking? Here's the solution: let your hair down every once in a while. After the party's over, compensate. For the rest of the week, eat lots of the nutritious, delicious, low-fat foods that should make up most of your regular diet:

✔ Fresh fruit (virtually no calories from fat)

✔ Salads (ditto – but watch that dressing!)

✔ Roast white meat (such as turkey), with no skin (20 percent of its calories from fat)

✔ Pasta (5 percent) with tomato-based sauces (2–3 percent)

By the end of the week, you're likely to have averaged out to a desirable amount with no problem and be right in line with that headline from the first page of the guidelines that we mention in the beginning of this chapter: 'Enjoy your food.' Amen to that.

A Few Words about Activity

When you take in more calories from food than you use up running your body systems (heart, lungs, brain and so forth) and doing a day's normal activity, you end up storing the extra calories as body fat. In other words, you gain weight. The reverse is also true. When you use more energy in a day than you take in as food, you release the extra energy you need out of stored fat and you lose weight.

We're not mathematicians, but we can reduce this principle to two simple equations in which E stands for energy (in calories), > stands for greater than, < stands for less than and W stands for weight:

E (in) > E (out) = +W

E (in) < E (out) = −W

It might not be rocket science, but you get the picture!

Calorie-burning in action

Chapter 6 has the lowdown on calculating the number of calories you can consume each day without piling on the pounds. Even being mildly active increases the number of calories you can wolf down without gaining weight. The more strenuous the activity, the more plentiful the calorie allowance. Suppose that you're a 25-year-old man who weighs 10 stone. The formula in Table 6-1 of Chapter 6 shows that you require 1,652 calories a day to run your body systems. Clearly, you need more calories for doing your daily physical work, simply moving around or exercising. Research shows that different levels of physical activity require different levels of energy or calorie intake. For example:

✔ Mild activity, such as gardening or housework, increases the number of calories you can consume each day without gaining weight to 2,645.

✔ Moderate activity, such as walking briskly at up to 4 miles per hour, raises the total number of calories you can consume each day without gaining weight to 2,810.

✔ Working full out at a heavy activity, such as playing football or digging, pushes the number of calories you can consume each day without gaining weight up to 3,471.

In other words, a 10-stone man who steps up his physical activity from mild to heavy can consume 826 extra calories without gaining weight. That happens to be just about the amount of calories in a normal serving of spaghetti bolognaise. Now we're talking!

If you've gained a lot of weight recently, have been overweight for a long time, haven't exercised in a while or have a chronic medical condition, you need to check with your doctor before starting any new exercise regime. (Caution: ditch any health club that puts you right on the treadmill without first checking your general health and fitness – heartbeat, respiration and so on.)

Finding activities you enjoy

The following list describes some forms of moderate activity for healthy adults as recommended by the *Chief Medical Officer's Report on Physical Activity and Health* (2004):

✔ Brisk walking (at the rate of 3 to 4 miles per hour)

✔ Cycling (at the rate of less than 10 miles per hour)

✔ Dancing

✔ DIY, such as painting the walls

✔ Gardening

✔ Golf

✔ Jogging

✔ Mowing the lawn

✔ Playing active games with your children

✔ Swimming

✔ Table tennis

Working out why you need to work out

Weight control is a good reason to step up your exercise level, but it isn't the only one. Here are four more:

- ✔ **Exercise increases muscles.** You can increase your muscle mass just by taking more regular exercise than you do at the moment. Because muscle tissue weighs more than fat tissue, some people may end up weighing more than they did before they started exercising to lose weight. But what your body weight is made up from is more important than the weight itself.

- ✔ **Exercise reduces the amount of fat stored in your body.** People who are fat around the middle as opposed to the hips (in other words an apple shape versus a pear shape) are at higher risk of weight-related illness. Exercise helps reduce abdominal fat and thus lowers your risk of weight-related diseases. Use a tape measure to identify your own body type by comparing your waistline to your hips (around the buttocks). If your waist (abdomen) is bigger, you're an apple. If your hips are bigger, you're a pear.

- ✔ **Exercise strengthens your bones.** Osteoporosis (a thinning of the bones that leads to repeated fractures) doesn't happen only to little old ladies. True, on average, a woman's bones thin faster and more dramatically than a man's, but after the mid-30s, everybody – male and female – begins losing bone density. Exercise can slow, halt or in some cases even reverse the process. In addition, being physically active develops muscles that help support bones. Stronger bones equal less risk of fracture, which, in turn, equals less risk of potentially fatal complications.

- ✔ **Exercise increases brainpower.** You know that aerobic exercise increases the flow of oxygen to the heart, but did you also know that it increases the flow of oxygen to the brain? When a heavy workload keeps you up working into the night, a gentle exercise break can keep you going. Dr Judith J. Wurtman is a nutrition research scientist at Massachusetts Institute of Technology and author of *Managing Your Mind and Mood Through Food*. She discovered that when you're awake and working during hours that you'd normally be asleep, your internal body rhythms tell your body to cool down, even though your brain is racing along. Simply standing up and stretching, walking around the room or doing a couple of sit-ups every hour or so speeds up your metabolism, warms up your muscles, increases your ability to stay awake and, in Dr Wurtman's words, 'prolongs your ability to work smart into the night'.

Chapter 14

Making Wise Food Choices

In This Chapter

▶ Using a food plate model

▶ Making sense of the nutrition information on food labels

▶ Choosing good food when you're shopping, cooking and snacking

*T*his chapter arms you with two valuable nutrition tools – a food plate model (known as the Eatwell Plate) and an understanding of food labels. These tools help you to make wise food choices.

Lots of facts and figures surround something as simple as deciding what to eat. But don't let multiple factoids and statistics turn you off. The information in this chapter is really handy for choosing the best food whether you're planning, shopping, cooking or snacking.

Introducing the Eatwell Plate

The Eatwell Plate (see Figure 14-1) is really a pictorial translation of the eight guidelines for a healthy diet that we outline in Chapter 13. The essential message of the Eatwell Plate is that you don't have to give up the foods you enjoy to be healthy – it's simply the balance you eat from the different food groups that counts.

The eatwell plate

Use the eatwell plate to help you get the balance right. It shows how much of what you eat should come from each food group.

Fruit and vegetables

Bread, rice, potatoes, pasta and other starchy foods

Meat, fish, eggs, beans, and other non-dairy sources of protein

Food and drinks high in fat and/or sugar

Milk and dairy foods

Figure 14-1: The Eatwell Plate shows the ideal quantities of different foods.

The eatwell plate shows how much of what you eat should come from each food group. This includes everything you eat during the day, including snacks.

So, try to eat:
* Plenty of fruit and vegetables
* plenty of bread, rice, potatoes, pasta and other starchy foods – choose wholegrain varieties when you can
* some milk and dairy foods
* some meat, fish, eggs, beans and other non-dairy sources of protein
* just a small amount of foods and drinks high in fat and/or sugar

Source: Adapted from the Food Standards Agency's eatwell plate (www.eatwell.gov.uk/healthydiet/eatwellplate).

Taking a close look at the Eatwell Plate

The Eatwell Plate is divided into sections of differing sizes representing the five common food groups:

- **Bread, rice, potatoes, pasta and other starchy foods:** Includes breakfast cereals, oats, corn, chapattis, yams and plantains.

- **Fruit and vegetables:** Includes fresh, frozen, tinned and dried varieties.

- **Milk and dairy foods:** Includes milk and calcium-fortified soya alternatives to milk, yogurt, cheese and fromage frais, but not butter or cream.

- **Meat, fish, eggs, beans and other non-dairy sources of protein:** Includes red meat, poultry, offal, fish, eggs and vegetarian sources of protein such as nuts, beans, pulses, tofu and Quorn.

- **Foods and drinks high in fat and/or sugar:** Includes spreading fats, oils, cream, salad dressings and sauces, cakes, pies, biscuits, pastries, sweets, chocolate, and savoury snacks such as crisps and soft drinks.

Bearing the groups in mind, the Eatwell Plate delivers two important practical messages:

- **Proportion matters.** The different sections of the plate aren't all the same size. The sections represent the proportions of the different food groups that make up a healthy diet. Eat lots of the first two groups (bread and cereals, and fruit and veg) – each segment of the plate represents about one third of your diet. Eat the next two groups – meat and alternatives, and dairy foods – in moderate amounts, about one eighth of your daily food intake as meat (or alternative) and one sixth as dairy.

  Fatty and sugary foods aren't essential to your diet but add choice and palatability. Consider them as occasional foods, comprising a much smaller part of your food intake – moderation is the key here.

- **Variety counts.** You need a variety of the foods represented in the Eatwell Plate to get all the nutrients you require. Choose different foods from within each group to increase your range of nutrients. The Eatwell Plate advises eating whole grain varieties from the bread and cereal group where possible, and eating a wide variety of different types of fruit and vegetables. Choose low-fat versions of the foods from the meat and dairy groups whenever you can. These include semi-skimmed or skimmed milk, low-fat yogurts and cheeses that are lower in fat than cheddar such as Edam, half-fat Cheddar or Camembert. Choose lean meat and mince, poultry without skin and fish without batter, and avoid frying where you can. Opt for lower-fat spreading fats, dressings, cakes, biscuits and ice cream where available.

Tailoring the Eatwell Plate to your needs

Different people have different energy (calorie) needs, which can be quite difficult to measure accurately. Healthy people with higher energy requirements need to eat larger portion sizes and more servings than those with lower energy expenditure. Nevertheless, the relative proportions of food from the different groups in the Eatwell Plate should stay the same.

How much food you need is affected by your:

✔ **Activity level:** The more active you are, the more energy you need.

✔ **Age:** As you get older and are no longer growing you need less food.

✔ **Gender:** Men generally need more food than women.

A young, active person may eat more than someone who's retired, but the balance of the diet – the types of food and the proportions – should still stay the same.

The beauty of the Eatwell Plate is that using it enables you to eat practically everything you like – as long as you follow the guidelines on variety and proportion. Your own specific energy needs then determine the quantity you eat.

 The principles of the Eatwell Plate apply to most adults of any ethnic origin, including vegetarians. The Eatwell Plate doesn't apply to children below the age of 2 (who need an energy-dense diet including full-fat dairy foods). However, you can start to encourage children aged 2 to 5 years to work towards the guidelines as they adopt a more adult diet.

Understanding Food Labels

You can use food labels along with the Eatwell Plate guide to help you select the best types of food from each group.

Once upon a time, the only reliable consumer information on a food label was the name of the food inside. Then came the ingredients list that tells you in descending order of weight what's inside the package. Now you can find more detailed nutritional information labels on the majority of foods. More than 80 per cent of UK-produced, pre-packaged foods now have nutritional information on the packaging. European Union legislation strictly defines the format, but manufacturers aren't legally obliged to supply nutritional information unless a food makes a particular nutritional claim. Extensive reviewing is underway to make providing nutritional information a legal requirement.

The nutritional panel on packaged foods lists the main nutrients in the food. Energy is listed as calories and kilojoules, and fat, protein and carbohydrate are listed in grams. Table 14-1 shows a typical nutritional food label listing the nutrients sometimes referred to as the 'Big Eight'. Sugars, saturated fat, fibre, sodium and sometimes salt are also given.

Nutritionists measure in 100 gram or millilitre portions, and so do nutritional labels. Labels also optionally list nutrients per serving.

Table 14-1	Typical Nutrition Information Panel Showing the 'Big Eight' Nutrients	
Tomato and Mushroom Pasta Sauce		
Typical Values	*Per 100g*	*Per ½ pot*
Energy	279 kJ/67 kcal	418 kJ/100 kcal
Protein	3.0g	4.5g
Carbohydrate		
(of which sugars)	6.0g	
(5.4g)	9.0g	
(8.1g)		
Fat		
(of which saturates)	3.4g	
(0.6g)	5.1g	
(0.9g)		
Fibre	1.4g	2.1g
Sodium		
(Salt)	0.4g	
(1.0g)	0.6g	
(1.5g)		

Some pre-packed foods contain high levels of energy, salt, fat, saturated fat or sugar, but to find this out you often have to study the back of pack quite carefully if you want to compare foods or select healthier options. Recent Food Standards Agency research showed that nearly half of all consumers now check labels for this key information and value having it in a more easily accessible format on the front of pack. As a result many retailers and manufacturers in the UK have developed 'nutritional signposting' schemes. At present no one agreed system is in use, so depending on where they're shopping, shoppers may be confronted by one of two main systems in use: traffic lights and guideline daily amounts.

Negotiating the traffic lights

Many food companies have adopted a colour-coded system for foods based on traffic lights, where a red (high), amber (moderate) or green (low) colour is given to the food for each of the four core nutrients – fat, saturated fat, added sugars and salt – per portion. These colours are based on agreed and standardised levels set by the Food Standards Agency. The colours are usually supported by a numerical value giving the actual amount of the nutrient per serving. This information doesn't replace full nutritional information on the back of the pack.

In research many consumers found this system easiest for comparing similar foods and for making a judgement about the level of a particular nutrient in a food. But some people have criticised the traffic light system as being too simplistic. These critics claim that red is too emotive, leading to the demonisation of certain foods, and that amber is confusing. Defendants of the scheme point out that some discrepancies are inevitable in any system but that this at-a-glance system works best for pre-packaged convenience items where large amounts of fat, salt and sugar could remain hidden from the consumer.

Checking out guideline daily amounts

Following extensive research on what consumers wanted to help them understand and interpret food labels better, the Institute of Grocery Distribution developed guideline daily amounts (GDAs). GDAs are a guide to the daily levels of energy, fat, saturated fat, sugar and salt in a healthy diet for adults and children (see Table 14-2). Actual needs vary according to age and activity levels, but GDAs are a useful benchmark. For most nutrients other than fibre,they represent upper ceilings rather than targets to aim for, but you can use GDAs to see how a particular food fits into a healthy diet.

Table 14-2	Guideline Daily Amounts for Adults and Children		
Nutrient	*Women*	*Men*	*Children (Aged 5—10 Years)*
Calories	2,000 kcals	2,500 kcals	1,800 kcals
Fat	70g	95g	70g
Saturated fat	20g	30g	20g
Carbohydrate	230g	300g	220g

Nutrient	Women	Men	Children (Aged 5—10 Years)
Protein	45g	55g	24g
Salt	6g	6g	4g
Fibre	24g	24g	15g

Source: www.gdalabel.org.uk

Many of the food companies that rejected the idea of traffic lights use as an alternativeschemes based on the percentages of the adult female GDAs provided by a serving of the food. In addition, a few products designed specifically for children contain information based on children's GDAs. Usually, the GDAs are monochrome or pastel coloured to avoid any emotive colour-coding.

However, the GDA system has limitations. When researchers showed consumers the GDA labels, some were turned off by the complex information. Research has also pinpointed difficulties in using GDA labels to identify whether a food has high, medium or low levels of nutrients, and in understanding what the percentage actually represents, and suggested consumers preferred a colour-coded system.

Given that studies show the average shopper buys 61 items in 26 minutes during a weekly shop, leaving just a few seconds to read each label, a lack of consistency between manufacturers and retailers and any potential for confusion is less than ideal. A clear format that's useful to the widest range of consumers is a better option, and one that's receiving attention in the EU at present.

Understanding Nutritional Claims on Food Labels

Sometimes manufacturers make specific claims about the nutrient composition of a food – for example, saying the food is high in a particular nutrient ('a good source of vitamin C') or that it contains more or less of the nutrient than a standard item ('low in fat' or 'high in fibre'). These nutrient claims are useful signposts for the busy consumer. And the good news is, nowadays the law imposes strict controls to protect consumers from false or misleading claims.

Regulations covering the agreed nutrient claims that can appear on food labels have recently come under a new EU regulation on nutrition and health claims. The EU are still standardising definitions, but we summarise the general consensus in Table 14-3.

Table 14-3	Guidelines for Common Nutritional Claims per 100 Grams or 100 Millilitres
Nutritional Claim	*Definition per 100g (100 ml)*
Low calorie	40 kcal or less (20 kcal for drinks)
Low sugar	5g or less (2.5g for liquids)
Low fat	3g or less (1.5g for liquids)
Low saturated fat	1.5g or less (0.75 g/100 ml)
Low sodium	120mg or less per 100g or 100 ml
Reduced sugar	Contains at least 30% less than standard product
Reduced fat	Contains at least 30% less than standard product
Reduced salt	Contains at least 25% less than standard product
Sugar free	0.5g or less per 100g or 100 ml
Fat free	0.5g or less per 100g or 100 ml
High fibre	6g per 100g (or 3g per 100 kcals)
Source of fibre	3g per 100g (or 1.5g per 100 kcals)

*Source: European Food Safety Health & Nutrition Claims (*http://ec.europa.eu/food*).*

In the past manufacturers sometimes added to the confusion by using terms such as 'extra lite' or a '% fat free' claim. This was particularly misleading because an 85 per cent fat-free food still contains a hefty 15 per cent fat. This practice is no longer permitted under the new regulations.

While still a better choice than standard crisps, a 'reduced-fat' bag of crisps with 30 per cent less fat than normal may still contain quite a lot of fat. Look at the nutritional information and compare it to your GDA or to the traffic light label on similar products to find the best option.

Using the Eatwell Plate and Food Labels in Practice

The Eatwell Plate and the information on food labels are easy to use when you become familiar with them. Here are some possibilities for using these tools to make shopping, meal planning, cooking and even snacking better – and better for you.

Healthy shopping

The Eatwell Plate proves that you don't need to avoid certain foods forever; you don't need to avoid certain aisles in the supermarket either. Think of your local shops or supermarket like the Eatwell Plate. Make a shopping list based on the five sections and try to stick to it – and avoid shopping when you're hungry or you'll be tempted to buy all sorts of extras from the fatty and sugary group as a quick fix. Visit all the sections and make sure that your trolley looks like the plate when you've finished!

You could even try dividing your allocated shopping time in proportion to the food model too. Spend time checking out the range of exotic or seasonal fruit and vegetables and the huge variety of starchy foods from couscous to polenta. By contrast, just nip in to the cakes, biscuits and puddings section – don't browse there all day!

Understanding food labels enables you to have your cake and eat it – nutritiously. At the supermarket, compare the labels of similar products and choose the best alternatives within each group. You gradually become aware of the healthier types from each group of foods – you can't do the whole trolley all at once.

You can also use nutritional labels to check for ever-increasing portion sizes on ready meals and snacks such as sandwiches, crisps and chocolate bars. Just compare the fat content of a standard-size chocolate bar over a king-size bar and we think you'll pick the smaller bar every time!

Healthy meal planning

Back home in the kitchen, when you start deciding what to eat or preparing meals, the Eatwell Plate can help you plan anything from a nutritious packed lunch to a roast dinner. Achieving the balance of all groups at every meal isn't always possible or practical, but over the day the Eatwell Plate can help to get things in proportion.

What about dishes made up of more than one food group? Most of the food you eat is made up of a combination of different food groups, such as casserole, pies, pizza, lasagne and even the humble sandwich. The trick here is to think about how the ingredients within the dish relate to the groups shown in the plate. Take the example of spaghetti bolognese. The pasta comes from the bread group; the beef from the meat and alternatives group; the Parmesan cheese from the dairy segment; the oil for frying from the fatty food group; and the tomatoes and onions in the sauce from the fruit and vegetables group. The healthy plate has a relatively larger serving of pasta with smaller servings of the meat and cheese. You can see quite easily that the proportion of vegetables in the dish doesn't stand up to the Eatwell Plate, but if you serve the meal with a side salad or add more vegetables to the sauce, your meal gets much nearer to the guidelines.

Healthy snacking

Our handy tools can even help you with health choices on the hoof. In the Eatwell Plate snacks as well as meals count towards the proportions in your diet. Missed out on fruit and vegetables at breakfast? Well, what could be a handier mid-morning snack than unzipping a banana or cracking open a kiwi? Carrot sticks, celery or mini cherry tomatoes are easy to eat. Dried fruit, such as a mini box of raisins, counts towards your intake of the fruit and veg group, as do individual tins of fruit – use the food label to find those with no added sugar.

Breadsticks, crumpets, muffins, crispbreads and oatcakes are alternatives to bread to fill that gap, and snacking on a bowl of cereal is another great way of increasing your starchy group. A bit low on your dairy group? Grab a smoothie made with fruit and milk, a pot of fruit yogurt, a fromage frais or even a low-fat rice pudding.

With items such as cakes, biscuits and savoury snacks, use the nutritional information on the food label to compare them and choose the healthier varieties from each group. To get you started, Table 14-4 gives you some alternatives to high-fat snacks.

Table 14-4	Snack Choices
High Fat Snack	*Lower Fat Choices*
Cakes	Scones, teacakes, malt loaf, currant buns, raisin bagel
Biscuits	Ginger nuts, Jaffa cakes, garibaldi, fig rolls
Savoury snacks	Plain popcorn, rice cakes, pretzels, rice crackers

Eating Well Despite the Credit Crunch

A major concern when you go food shopping may be cost rather than nutrition, but the good news is these can go hand in hand.

Here are some tips for savvy food shopping:

- ✔ Buy what you *need* first and what you *want* for treats later – think beans before biscuits, cereal before crisps and chocolate, and lean meats before muffins!

- ✔ Loose packed items including fruit, vegetables, herbs, spices and cereals are often better value for money.

- ✔ Buy fresh foods in season when they're cheaper, and avoid pre-washed, pre-prepared fruit and vegetables. Frozen and canned fruit and vegetables can be good value. And why not try growing fresh herbs in pots on the windowsill?

- ✔ Swap premium brands for budget labels. Remember: 'eye level is buy level' where the shop places the most profitable items. Look up or down on the shelves for cheaper alternatives.

- ✔ Budget stores or supermarket economy ranges are often more economical and just as nutritious, especially for staple items such as bread, beans, yogurt, vegetable oil, meat, fish ,eggs, pasta, oats and other cereals and juices. Economy range, uneven-shaped fruits and vegetables are much cheaper and just as nutritious.

- ✔ Shop with a friend to make the most of bulk buy savings or multiple special buys of fresh produce. Or make use of using special deals on foods that can be frozen (e.g. meat, fish, chicken or bread) or those with a long shelf life e.g. canned goods.

- ✔ Try to eat more vegetarian meals based on using nuts, eggs or beans and legumes.

- ✔ Reduce food waste. Not only will you save money, but you'll help the environment too (see Chapter 20 for more on sustainability).

Many people are heavily influenced by the 'buy one, get one free' (BOGOF) special offers that are very often on foods with higher levels of fat, salt and sugar. Research estimates around one in four people buys into BOGOF offers because they believe they're saving money. However, these offers can be a false economy if you don't really need the extra products, and they can be bad for your diet if they stop you buying more healthy foods. Watch out for shops promoting BOGOF offers at eye level or at the end of aisles where traffic is heavy and slower, and think before you pop foods on offer in your trolley.

At the beginning of this chapter we warned you that keeping track of all the facts and figures might be difficult. But you can pretty much sum everything we say up in one golden rule exemplified by the Eatwell Plate (we could even write it on the food label):

Keep things in proportion.

Come to think of it, that's not a bad philosophy for life.

Chapter 15

Ensuring Good Nutrition Whoever You Are

A good diet is one that provides sufficient amounts of all the essential nutrients your body needs. But how do you know what you need, and how do your needs vary according to your stage in life?

Homing In on the Dietary Reference Values

In 1991 the Department of Health produced a comprehensive guide to essential nutrients, in the light of new research and the best available evidence from various sources, including both animal and human studies and those on the relationship between diet and health in groups of people (*epidemiological* studies). The guide, entitled 'Dietary Reference Values (DRVs) for Food Energy and Nutrients for the UK remains the UK nutrition bible. DRVs give a good yardstick by which to compare the intake of different groups in the population and help to identify those at risk.

DRVs aren't exact nutrient requirements for individuals but they are sound guidelines.

The four types of DRVs are:

- **Estimated average requirements (EARs):** The estimated mean or average requirement of a group for a particular nutrient. This is the DRV used for energy. If you exceed your need for energy you end up with excess calories, which may cause you to put on weight. The DRVs for fat and carbohydrates are expressed in averages as a percentage of your food energy, because the proportions in your diet are what matter.

- **Reference nutrient intakes (RNIs):** The amount of a nutrient deemed sufficient for almost all healthy individuals (97.5 per cent), in fact more than most people need. RNIs are used for protein, and for 9 vitamins and 11 minerals.

- **Lower reference nutrient intakes (LRNIs):** The amount of a nutrient deemed sufficient for only a few individuals in the population (2.5 per cent). Regular intakes below the LRNI are almost certainly inadequate.

- **Safe intakes:** These are used where not enough evidence exists to set an EAR, RNI or LRNI. Safe intakes are set to meet the needs of most people while keeping below the level at which they could have toxic effects. Safe intakes are currently set for four vitamins and four minerals.

This chapter homes in on the specifics of how these requirements may vary. We take you through the lifecycle – visiting the different ages, sexes and physiological states such as pregnancy to explore the key nutritional concerns for each group.

Remember that the dietary recommendations for different ages are developed on the best available evidence at the time and are updated – or reviewed, revised or rehashed – in the light of new information. Dietary recommendations are rarely set in stone!

I'm Trying for a Baby

A healthy balanced diet is important at any stage in a woman's life. Women's bodies face great physiological challenges, and a balanced diet is never more important than when a woman is preparing her body for pregnancy.

Body fat has an important influence on female fertility: women who are either very underweight or very overweight may find it more difficult to conceive and support a pregnancy to full term. This isn't the time for crash diets, however. During the early stages of pregnancy a foetus is very susceptible to nutritional imbalances and a very low-calorie, extreme diet is almost certainly not going to give you sufficient quantities of nutrients for you to build up a good store to support both you and a growing baby.

Reviewing terms used to describe nutrient recommendations

DRVs for energy, fat and carbohydrate are given either in calories or as a percentage of total food energy. The RNIs for protein are given in grams, and those for vitamins and minerals are in milligrams (mg) or micrograms (mcg). A milligram is ⅟₁₀₀ of a gram; a microgram is ⅟₁₀₀ of a milligram.

Vitamins A, D and E are special cases:

✔ One form of vitamin A is preformed, which means that your body can use it right away. Preformed vitamin A, known as *retinol*, is found in food from animals – liver, milk and eggs. Carotenoids (yellow pigments in plants) also provide vitamin A. But to get vitamin A from carotenoids, your body has to convert the pigments to chemicals similar to retinol. Because retinol is a ready-made nutrient, the RDA for vitamin A is listed in units called *retinol equivalents*

(RE). An RE is equal to one microgram of preformed vitamin A.

✔ Vitamin D consists of three compounds: vitamin D1, vitamin D2 and vitamin D3. Cholecalciferol, the chemical name for vitamin D3, is the most active of the three, so the RDA for vitamin D is measured in equivalents of cholecalciferol.

✔ Your body gets vitamin E from two classes of chemicals in food: tocopherols and tocotrienols. The compound with the greatest vitamin E activity is a tocopherol: *alpha*-tocopherol. The RDA for vitamin E is measured in milligrams of *alpha*-tocopherol equivalents (mg a-TE).

Don't worry – we won't test you on these. Head over to Chapter 9 for the lowdown on vitamins.

The Department of Health recommends that women who are trying for a baby should start taking a 400-microgram supplement of folic acid as soon as they stop using contraception, right up until the 12th week of pregnancy. Folic acid can reduce the risk of the baby developing a neural tube defect such as spina bifida. If your pregnancy is unplanned, take folic acid as soon as you can and continue until week 12. You can get dietary sources of folate from fortified bread and breakfast cereals and dark green leafy vegetables. Eat these in addition to a supplement.

It's a good idea to build up your iron stores leading up to a pregnancy because iron helps your cells to make haemoglobin (red blood cells) for both your own body and your baby's. The DRV for iron doesn't increase during pregnancy, but this is because your periods usually stop (so you don't lose iron every month), your body adapts to pregnancy and you can absorb more iron from food; you're also able to mobilise iron stores laid down before pregnancy. Most women probably don't need to take an iron supplement before they become pregnant, but eating at least one rich iron source every day is a good idea. Foods such as lean red meat, pulses and beans, green vegetables and fortified breakfast cereals are all good iron supplies, and drinking some vitamin C-rich orange juice at the same time helps improve absorption.

Avoid high-dose vitamin A supplements when you're pregnant, and multivitamin supplements not specifically designed for pregnant women. Very high intakes of vitamin A before and during pregnancy are associated with an increased risk of miscarriage and congenital malformations. You get all the vitamin A you need from a varied diet. Avoid taking vitamin A supplements or eating liver and liver products like pâté. Just like humans, animals store their vitamin A in the liver, which means it can be a particularly concentrated source.

If you're planning to have a baby, it's advisable to limit your intake of certain types of fish to minimise your risk of absorbing mercury. Mercury can be harmful to a baby's nervous system. Avoid eating shark, swordfish, marlin and tuna. Your intake of such fish shouldn't exceed 140 grams cooked weight per week. That's the equivalent of one fresh tuna steak per week or two medium-sized cans of tuna.

Although you don't need to cut out alcohol altogether before you become pregnant, it's advisable to drink no more than 1 or 2 units once or twice a week. (A unit is half a pint of beer, lager or cider, a small glass of wine or a single 25-millilitre measure of spirits.)

I'm Pregnant

Aim to continue the recommendations in 'I'm Trying for a Baby' throughout your pregnancy. Stop folic acid supplements after the 12th week, and ideally avoid alcohol until the final few weeks of pregnancy when one or two units once or twice a week is okay, although of course if you can avoid drinking any alcohol that's the best policy.

Try to avoid too much caffeine during your pregnancy. High levels of caffeine can lead to low birth weight or even miscarriage. The recommended safe limit of caffeine is 300 milligrams a day. Each of these servings contains roughly 300 milligrams caffeine:

- ✔ Three mugs of instant coffee (100 milligrams each)
- ✔ Four cups of instant coffee (75 milligrams each)
- ✔ Three cups of fresh brewed coffee (100 milligrams each)
- ✔ Six cups of tea (50 milligrams each)
- ✔ Eight cans of cola (up to 40 milligrams each)
- ✔ Four cans of 'energy drink' (up to 80 milligrams each)
- ✔ Eight 50-gram bars of chocolate (up to 50 milligrams each)

When Baby wants a snack . . .

It's quite normal to gain between 10 and 12 kilograms in weight over the course of a pregnancy, but you may find that you gain considerably more than this. Your appetite can increase during pregnancy and you feel you want to snack more. The best foods to snack on are fruit; sandwiches filled with lean meats, cottage cheese or chicken; and low-fat yogurts.

You may find that your favourite food becomes completely intolerable when you're pregnant.

This is very common and usually the particular food becomes your favourite again after giving birth. On the other hand, you may experience cravings for foods that you'd normally consider inedible. Women have craved laundry starch, rocks, matchboxes and clay before! Don't worry – no evidence exists that this phenomenon (called *pica*) indicates a mineral deficiency or in fact has any physiological significance.

The old adage that a pregnant woman is eating for two suggests that she has dramatically higher nutrient needs. With a few key exceptions (folate and vitamin D), most women get everything they need for themselves and the baby from a normal well-balanced diet. Your energy, protein and some vitamin needs are slightly higher than usual, but nothing too drastic. Some women may benefit from taking a vitamin D supplement to achieve the increased requirement of 10 micrograms per day. Margarines, cheese, fatty fish and eggs usually supply enough vitamin D, along with exposure to sunlight, but women who don't get enough exposure to the sun for cultural or religious reasons may find supplements particularly useful.

Food hygiene is particularly important during pregnancy. Infections can affect both you and the developing child, so take special care in the kitchen and beyond.

Avoid foods during your pregnancy that can contain high levels of listeria, a germ that can cause miscarriage, stillbirth or severe illness in a newborn baby. Give the following foods a miss:

- **Soft mould-ripened cheese, such as Camembert, Brie and blue-veined cheese.** Hard cheeses like Cheddar and soft cheeses such as cottage cheese or processed cheese are safe.

- **Pâté.** Avoid all types of pâté including vegetarian pâté.

- **Uncooked or under-cooked ready meals.** Make sure that you heat ready meals until they're piping hot all the way through.

Other foods to take extra care with:

- Avoid eating raw eggs and food containing raw or partially cooked eggs. Only eat eggs where both the white and the yolk are solid. This is to avoid the risk of salmonella, which causes a type of food poisoning.

> ✔ Always wash your hands after handling raw meat, and keep raw foods separate from ready-to-eat foods. This is to avoid food poisoning germs such as salmonella, campylobacter and E. coli 0157.
>
> ✔ Make sure that you only eat thoroughly cooked meat. Be particularly careful with sausages, minced meats and barbecued meats.

Always wear gloves when gardening or emptying cat litter trays and wash your hands thoroughly afterwards. This is to avoid *toxoplasmosis*, an infection caused by a parasite found in meat, cat faeces and soil. This applies whether you're pregnant or not!

I'm Breastfeeding

Breast milk provides all the nutrients a baby needs for healthy development in the first months of life. Most women's bodies are very efficient at making breast milk, so you don't need to 'eat for two'. However, your requirements for energy, protein and most vitamins and minerals are higher than normal. So just like any other time, it's important for you and your baby that you eat a healthy, balanced and varied diet. While you're breastfeeding you need to continue to supplement your diet with 10 micrograms of vitamin D each day. If you receive Income Support or Jobseekers' Allowance you're entitled to vitamin A, C and D supplements.

It can be difficult to find the time to eat properly when you're looking after a new baby, so remember to keep meals simple. Try to schedule in time to eat; small, frequent meals may be easier to manage than two or three large meals a day.

You can eat most foods that you avoided during pregnancy without a problem. However, limit shark, swordfish, marlin and tuna to one 140-gram fish steak or two medium tins of tuna a week. This is because these types of fish can contain low levels of pollutants that can build up in the body over time, including dioxins and PCBs (Polychlorinated bephenyls).

If you have a family history of food allergy or intolerance, it may affect your baby. Avoid known allergens to keep on the safe side. Women were once told to always avoid peanuts while breastfeeding to reduce the risk of their baby developing an allergy to peanuts. This advice was changed in 2009 when a review of the evidence showed that it made no difference whether a mother ate peanuts or not. Also keep in mind that it's normal for breastfed babies to have loose stools, and this is unlikely to be a sign of a food allergy in your baby.

It's important to make sure that you drink regularly while breastfeeding. If your urine is dark and has a strong smell, it means that you're not drinking enough. Get in the habit of having a drink with you when you settle down to feed your baby. The best drinks are water, milk or unsweetened fruit juice. Small amounts of what you eat and drink may pass to your baby through breast milk, so limit your alcohol and caffeine intake.

I'm Weaning My Baby

During the first year a baby grows more quickly than at any other time. This rapid period of growth means that babies need a lot of nutrients to ensure that they can grow well. Initially, babies only need breast milk, or a suitable formula milk, but as they get older they need to have other sources of nutrition to help with growth and development.

Weaning is the introduction of solids into the diet of a baby who's drinking breast or formula milk. Children need weaning to get all the nutrients they require from the foods they eat. Weaning also helps with other aspects of development such as biting, chewing and eventually speech.

The Department of Health recommends exclusive breastfeeding until the age of 6 months (26 weeks). All infants, breastfed and formula fed, should be weaned at 6 months. Some parents prefer to wean earlier, but four months (17 weeks) is the very earliest that weaning onto solids should start. Babies who are born pre-term need to be weaned according to their individual needs; seek advice from your dietitian and medical team.

At around 6 months old a child's stores of some nutrients such as iron start to run out. So it's important that when you introduce cow's milk as the main drink, children are eating a varied diet that meets their nutritional requirements. Weaning foods are often introduced in stages. Table 15-1 shows the different stages of weaning and the sorts of foods to introduce.

The ages in Table 15-1 are approximate and depend on when weaning starts. Every baby is different and develops at his or her own pace. But you need to keep offering different tastes and textures through the first year.

Table 15-1		Weaning Stages	
Stage	*Age Range*	*Consistency*	*Foods*
Stage 1	6 months (26 weeks). No earlier than 4 months (17 weeks).	Smooth puréed food.	Fruit, vegetables, rice, potatoes, yam, meat, yogurt, cheese and custard. (Before 6 months foods should be gluten free.)
Stage 2	6–9 months.	Thicker consistency with some lumps; you can also introduce soft finger foods at this stage.	As above, but you can now start to introduce bread, cereals, pulses and eggs.

(continued)

Table 15-1 *(continued)*

Stage	Age Range	Consistency	Foods
Stage 3	9–12 months.	Mashed, chopped, minced consistency; more finger foods.	As above.
Stage 4	12 months and over.	Mashed, chopped family foods and a variety of finger foods.	As above.

Source: Based on the Paediatric Group of the British Dietetic Association position statement on breastfeeding and weaning onto solid foods.

Breastfed babies don't need additional drinks, but formula-fed babies may need some extra water (cooled boiled water) in hot weather. Don't give your baby fruit juices, because even natural sugars can cause tooth decay. After 6 months, still regularly breastfeed your baby at mealtimes. If your baby is formula fed, give him or her 500–600 millilitres of suitable infant formula milk once a day.

 It's best to wean children onto the foods that the family eat. Although be careful not to add salt to your family recipes. Children who only eat commercial baby foods may not like family foods when you offer them. However, you can of course include some commercial baby foods in the weaning diet; many parents find them very convenient.

Baby led weaning

There is a growing trend for baby led weaning (BLW) instead of the traditional route of starting with purees then moving on to lumpy foods and finger foods. This ties in with the guidelines that suggest that weaning should start from around 6 months. Baby led weaning is still not recommended by a lot of health professionals because there is concern that babies may not receive all the nutrients they need.

In theory baby led weaning, like breast feeding, allows babies to learn appetite control, so they eat when hungry and stop when full. This may help reduce the chance of obesity later in life.

BLW involves offering babies a range of foods and allowing them to explore and select their foods and eventually self feed. It can be seen as a natural extension from breast feeding, but bottle fed babies can be weaned this way too.

Initially your baby might only touch and play with the food, before moving on to licking, tasting and finally eating some. When offering foods such as rice and cereals you can give your baby a spoon, but they will probably start eating it with their fingers first before mastering the use of a spoon. Be warned, this can be a messy process, so you might want to invest in a messy mat and some bibs first!

My baby is vegetarian or vegan

Children can grow and develop normally on a vegetarian or vegan diet, although you need to ensure that your baby's diet meets his or her nutritional needs. Vegetarian and vegan diets can be high in fibre, and this can lead to low energy (calorie) intake and reduced absorption of some important minerals, such as iron and zinc. All children between 6 months and 5 years old can benefit from taking vitamin drops containing vitamins A, C and D, but vegan children additionally need vitamin B12. A health visitor or dietitian can give you specific advice on weaning onto a vegetarian or vegan diet.

Foods to avoid giving your baby

A couple of foods are a no-no when it comes to feeding your baby:

- ✔ **Salt:** Up to 7 months, give your baby less than 1 gram of salt per day. (Both breast milk and formula milk contain the right amount of salt.) Between 7 months and 1 year, 1 gram of salt per day is the maximum. Don't add salt to foods for your baby, and limit salty foods like bacon, cheese and some processed foods.

- ✔ **Sugar:** Avoid adding sugar to foods and drinks for babies. Don't give honey (even for easing coughs) to your baby until they are a year old. Very occasionally honey contains a type of bacteria that can produce toxins in babies' intestines. This can cause a very serious illness called infant botulism. Honey is also a sugar, which means, like sugar, it can encourage a sweet tooth and potentially lead to future tooth decay.

I'm Feeding My Toddler

Just like adults, young children need food for energy, as well as for nutrients such as protein, fat, carbohydrates, vitamins and minerals. This is to make sure that their bodies work properly and can repair themselves.

Toddlers grow very quickly and are usually very active, so they need plenty of calories and nutrients. Just like adults, toddlers should be able to get everything they need from a varied, healthy diet.

Children can eat the same food as adults, but before the age of 2 they can't eat large amounts at one sitting. So until then, it's especially important to give them meals and snacks packed with calories and nutrients such as meat, full-fat milk and dairy foods and eggs, along with fruit, vegetables and starchy foods like bread, pasta, cereals, rice and potatoes.

Your toddler's stomach can't cope with too many high-fibre foods such as wholemeal pasta and brown rice. Too much fibre can reduce the amount of minerals absorbed such as calcium and iron. The best advice to is give toddlers a mixture of wholegrain foods and refined carbohydrate foods like white pasta.

By the age of 5 young children should be eating family food along with everyone else in the house. Family meals tend to be bulky, containing lots of starchy foods and plenty of fruit and vegetables. Make sure your toddler's meals don't contain too much saturated fat, found in butter, hard-fat spreads, cheese, meat and meat products, biscuits, pastry and cakes.

If your toddler is eating well and getting plenty of calories and nutrients from a varied diet, you can start giving him or her semi-skimmed milk (the most commonly consumed milk in the UK). Fully skimmed milk isn't suitable as a main drink until a child is 5 years old because it doesn't contain enough calories or vitamins.

My toddler is vegetarian or vegan

If you give your toddler a vegetarian diet, make sure it's balanced and includes foods rich in nutrients such as milk, cheese and eggs. This ensures that the diet isn't too bulky and contains plenty of protein, vitamin A, calcium and zinc. Vegan diets can make it very difficult to achieve a growing child's nutritional requirements; a paediatric dietitian can give you advice.

The iron found in meat is easier to absorb than other sources, so it's important to give your vegetarian or vegan toddler foods containing iron every day to avoid deficiency. Iron is found in many vegetables and pulses such as beans, lentils and chickpeas; in dried fruit, such as apricots, raisins and sultanas; and in fortified breakfast cereals. Give your toddler a drink high in vitamin C, such as fruit or vegetable juice, at the same time as the iron-rich food, to help with absorption. Don't give young children tea or coffee, especially at mealtimes, because this reduces the ability to absorb iron.

Foods to avoid giving your toddler

Toddlers eat sand, worms and homemade witches' potions when you're not looking. But limit the damage from food by following these recommendations:

✔ Don't give your toddler raw eggs or foods containing raw or partially cooked eggs because of the risk of salmonella poisoning. Make sure that both the white and the yolk of the egg are solid.

✔ Don't give whole nuts to children under 5 because of the risk of choking. Crush or flake them instead.

✔ Avoid giving fresh shark, swordfish or marlin to young children because of the relatively high levels of mercury, which can affect children's developing nervous system.

✔ Don't add salt, sugar or honey to food for your toddler. The tastes and preferences toddlers develop at this stage are going to be with them for a lifetime, so it's best to start with good habits now.

✔ Try not to give your toddler sweet fizzy drinks and fruit squash because they can lead to tooth decay. These drinks can also be filling, so your toddler doesn't eat enough food to get the nutrients he or she needs. If you do give fruit squash or sugary drinks, make sure that you dilute them well – at least five parts water to one part squash. Keep these drinks to mealtimes and stick to water or milk as between-meal drinks.

I'm Feeding My School Child

A child going through a major growth spurt needs a lot of body building blocks. As a result, the requirements for some nutrients are set particularly high for school-aged children and teenagers and a healthy diet becomes especially crucial. But wouldn't you just know it – this is the time when your children rebel against all your good parenting practices and develop their own idiosyncratic food choices.

Surveys of what UK school children are eat, such as the National Diet and Nutrition Survey (NDNS) of Young People published in 2003, make dismal reading. The intake of many vitamins and minerals is frighteningly low compared to the reference nutrient intakes (Chapter 13 has more on RNIs).

Look out! School dinner's about

Children's tastes develop very early in life; foods they learn to like and dislike as children both at home and school can stay with them for a lifetime. Countless studies show that what children eat during the school day can have a significant effect on children's behaviour and concentration in the classroom. So perhaps it's time that food and nutrition took a higher profile on the curriculum as well as in the dining hall.

Your bones accumulate in density and eventually reach the strongest point they'll ever be – the *peak bone mass* – in your late teens and early 20s. Having adequate calcium and vitamin D is especially important to get good bone strength in the first place and then reduce the risks of osteoporosis in later life. Calcium forms the matrix of the bone and vitamin D is used to absorb it. The NDNS showed low levels of both calcium and vitamin D in a proportion of school-aged girls and boys. Unfortunately, as they get older, children often go off the best sources of calcium such as dairy foods. Girls in particular often cut out all milk in a mistaken concern over body weight and fat intake.

Getting your children to drink milkshakes or yoghurt-based smoothies made from semi-skimmed milk is an excellent way to boost calcium intake.

School children need plenty of iron for making blood and lean muscle. Teenage girls are at risk of anaemia if they haven't built up good stores prior to menstruation. The NDNS found intakes of iron below the RNI in girls, and low body iron stores in both sexes. Ten per cent of 15 to 18-year-old girls are vegetarian, reducing iron intake even more. Try to ensure a good alternative vegetarian source of iron such as pulses, green leafy vegetables, dried fruit or enriched cereals (along with vitamin C to help absorb it).

A fortified breakfast cereal is a great way to start the day. It can help boost performance in lessons and give the fuel for sport. Vitamin C from fruit or fruit juice is essential for healthy skin. Make a good diet more attractive to your kids by promoting the effects of diet on appearance, body image, mood and even how well they do at school.

Also take inspiration from the recent move towards improving nutrition in school children. The TV chef Jamie Oliver has led a high profile campaign to improve the nutritional standards of foods served in schools. In 2005 the School Food Trust was established. And in 2006 the government introduced nutrition standards for food served in primary and secondary schools (see Chapter 17).

I'm a Man

Several dietary factors are of particular importance for men. A recent National Diet and Nutrition Survey (NDNS) showed that men eat more fats, oils, meat and meat products, and fewer fruit and vegetables and foods rich in fibre, than women. The most recent Health Survey for England confirmed that the average man's diet is too high in saturated fat and sodium, with only 27 per cent of men meeting the recommendations for five portion of fruit or veg a day. These dietary factors are all linked to an increased risk of heart disease and cancer. Studies also show that average levels of 'bad' (LDL) cho-lesterol in the blood are higher in men than women and that heart disease currently kills one in five men in the UK.

You can achieve a better balance by:

✔ Cutting down on fatty foods such as pies and pasties

✔ Eating more fruit and vegetables

✔ Ensuring starchy foods, such as bread, potatoes, pasta, rice and cereals, make up a third of your diet

✔ Eating oily fish once or twice a week to protect against heart disease

The Health Survey for England also found that 65 per cent of men were either overweight or obese – a greater proportion than for women. In addition, 31 per cent of the men had high blood pressure, putting them at increased risk of both stroke and heart disease, and 42 per cent regularly drank more than the safe levels of alcohol. High salt and alcohol intakes are both linked to high blood pressure, especially the increase in blood pressure that occurs as men get older. The average salt intake for men is 10 grams per day (1½ teaspoons) – but it really shouldn't exceed 6 grams (1 teaspoon). So be sure to cut down on salt added during cooking and at the table, and be aware of the salt content in processed foods, choosing lower salt varieties where possible.

Some men are at increased risk of heart disease and diabetes if they gain excess weight around their middles in the central abdominal area – the classic apple shape. As you get older or become less active, you may to eat need less food to avoid weight gain, while still being sure to get a good balance and variety from the main food groups.

I'm an Older Person

For some people getting older means not eating as well. However, good food is even more important as you get older and a poor diet can make you more susceptible to infections, slower to recover from injury or surgery and can affect your mood and quality of life.

Here are some factors that can cause older people to struggle to eat a healthy, balanced diet:

✔ Being ill, inactive, on medication or in pain can reduce appetite.

✔ Disability can impede shopping and cooking.

✔ Factors such as finances, lack of motivation, loneliness and depression can play a part in determining food intake and lead to missed meals.

✔ Lack of knowledge about food preparation – in a recent Health Survey for England 45 per cent of older men said they felt unsure how to shop and cook for a balanced diet.

✔ Loss of teeth, ill-fitting dentures and swallowing problems can make it hard to eat.

Malnutrition isn't a third world problem – it's surprisingly common in older adults, especially in residential care homes and hospitals, and even those who live in their own homes or sheltered accommodation. Nutrient deficiencies such as weight loss, anaemia or scurvy are quite common in older adults.

Here are some key areas of the diet to address:

- **Calcium:** Your body may have stopped growing by this stage in life but your tissues are still turning over. A daily calcium intake (from two to three servings of dairy foods) remains important for healthy bones.

- **Fibre:** Older adults often have fibre intakes well below the daily reference value (DRV) of 24 grams per day (surveys show that only around two out of three people have adequate intakes). Fibre helps reduce constipation and diverticular disease, both common in older adults. Wholegrain cereals, fruit, vegetables, nuts and pulses are all good fibre providers. Don't forget to drink plenty of fluid (at least six to eight cups a day) to ensure that the fibre does its job properly.

- **Folate:** This may play a role in helping to prevent heart disease, depression, osteoporosis and dementia, all of which can be age-related conditions. Good sources of folic acid include fortified and wholegrain cereals, beans and several fruits and vegetables. If chewing is a problem, tinned and stewed fruit and vegetables or a glass of juice or smoothie count towards your five a day.

- **Iron:** Good natural sources of iron (red meat, offal, pulses, oily fish, eggs) or fortified products (such as breakfast cereal) are important along with vitamin C to help absorb it. Where possible, avoid drinking tea with a meal, because the tannins can impair iron absorption.

- **Vitamin D:** This is essential to allow your body to absorb calcium and to help prevent brittle bones or osteoporosis, which leads to bone fractures. Many older people get insufficient exposure to sunlight to make their own supply of vitamin D and ageing reduces the ability to make it in the body. The recommended daily intake of vitamin D for people over 65 is 10 micrograms a day. Good dietary sources of vitamin D include oily fish, meat, fortified cereals, evaporated milk and margarine. A daily supplement is useful for many older adults. Ask your GP for advice.

Age UK has some other useful information on healthy eating for older adults, including the use of supplements, heart health, digestive problems, boosting your immunity and healthy bones. Check out `www.ageuk.org.uk`. And if you're interested in practical and nutritional guidelines for food in residential or nursing homes or for community meals, you can download for free a fantastic booklet called 'Eating Well for Older People' from the Caroline Walker Trust website at `www.cwt.org.uk/pdfs/OlderPeople.pdf`. See also the information in Chapter 17 on nutrition in residential care homes.

Chapter 16

Being Nutritionally Savvy for When You're Out and About

*T*he food you eat outside the home makes up an increasingly important part of your diet. The average person eats one in every six meals out of home, and if you add in snacks and grab-and-go food then the figure rises. Men consume about a quarter of their calories when eating out, and women around a fifth. So the choices you make when eating out can go a long way to helping you maintain a balanced diet.

The Food Standards Agency is working in a number of areas to help people make healthier choices when they eat out – whether this is in restaurants, pubs, cafes, service stations, at work and or at home with a takeaway.

In this chapter we look at initiatives to help make it easier to get hold of healthy food while you're at work. Whether you cruise the canteen, pack your own or provide the workers with their lunch, there's something here to help you make some smart choices. Heading out on a journey has nearly always meant one thing when it comes to food – greasy, sugary and down-right bad for you. So we guide you through the foods you can take with you to keep you alert while tickling your taste buds, and we also bring you up to date with what's available at the service station. Finally, eating out is just part of people's lives these days – you don't need a special occasion to go out for dinner and there's more choice than ever. Here you find the lowdown on menu information to keep things on the healthier side and also what to pick when no nutrition information is available.

Being Fit for Business: Nutrition in the Workplace

If you work somewhere that has a staff restaurant or canteen you may have seen a few simple changes over the past couple of years. You might find:

- ✔ A choice of spreads
- ✔ A fresh fruit bar
- ✔ A new salad bar
- ✔ Calories and other nutritional information on the menu
- ✔ Low fat options
- ✔ Water on the table

If you've seen things like this creeping into the canteen, chances are your company is taking part in a national drive to get folks at work eating well and taking more physical activity. The government started the campaign, and the Food Standards Agency is working with caterers to help them plan healthier menus and adapt recipes to save on calories, saturated fat sugar and salt.

So why would your employer worry about what you eat at work? Surely employers have other things to worry about, right?

Promoting healthy eating is part of promoting general health in the workplace, and research has shown that improving health in the workplace:

- ✔ Increases motivation and creates a better working atmosphere
- ✔ Increases quality of products and services
- ✔ Improves the public image of a company

Among other things poor nutrition is also associated with obesity, and with 260,000 working days lost a year as a result of obesity-related illness at a cost of £500 million, it's little wonder your boss wants you to eat well!

It's not just large companies with swanky canteens that can make a difference – even small organisations with just a staff room can make a few simple changes to encourage employees to bring their own healthy food in for lunch rather than making a bolt to the burger bar. Providing crockery and utensils, dishwashing and hand-washing facilities, a microwave oven and drinking water encourage employees to bring a packed lunch in from home rather than rely on fast food.

Telling your employer what you want

It's easy to complain about the food in your staff canteen, but what if someone actually listens and then asks you what you'd like instead! What would you say? Here are some ideas to help get you started. You might not get them all but you'll never know unless you try:

- **Types of food provided:** Offer a wide choice of fruit, vegetables and breads. Main ingredients should include rice, potatoes, breads and pasta. Offer fish and chicken more often, and include a variety of vegetarian options that include eggs, lentils, beans and quorn.

- **Ingredients:** Use a variety of fresh, frozen and tinned fruit and vegetables. Where possible use low fat products like semi-skimmed milk, low fat spread and low fat yoghurts. Reduce the amount of salt in cooking and remove it from tables. Use oils and sugar sparingly.

- **Food preparation:** Trim visible fat from meat, and skim fat from soup, stock and mince. Avoid preparing vegetables far in advance, over-cooking or storing for a long time before serving. Use grilling, steaming, boiling, casseroling, poaching, dry roasting and stir-frying; avoid deep fat frying wherever possible.

- **Display and promote:** Ensure salad bars, fresh fruit and a selection of breads are prominent in the serving area. Make baked potatoes more visible than chips, and offer salad dressings separately. Feature details and nutritional information of new recipes on the menu. Ensure healthy options aren't more expensive than unhealthy ones!

I'm an employer – what can I do?

Plenty of help is available for employers who want to encourage their staff to eat well. These steps are a good place to start:

1. **Consult with an outside specialist – a nutritionist or dietitian.**

2. **Set up a working group.**

3. **Consult, communicate and involve everyone in your organisation.**

4. **Establish what you currently do, and then develop an action plan for the way forward.**

5. **Implement, monitor and review your plans.**

Here are some useful places to find information about healthy eating at work:

✔ British Nutrition Foundation (www.nutrition.org.uk)

✔ The British Dietetic Association (www.bda.uk.com)

✔ The Food Safety Promotion Board (www.safefoodonline.com)

✔ The Food Standards Agency (www.food.gov.uk/healthiereating)

✔ Weight Wise (www.bdaweightwise.com)

Healthy eating at work doesn't have to mean big changes, but it can bring big rewards for staff and employers.

Grabbing food at your desk

One in five people prefer to take their own packed lunch to work, and this figure is likely to increase further as the economic climate gets colder! Packed lunches can get repetitive and boring, so we've put together some tips to keep them delicious and nutritious. The following sections help you make a healthy packed lunch that'll keep you satisfied and provide the balance you need to be healthy.

Getting the basics right

Keep a selection of different breads in the freezer for sandwiches – take advantage of reduced-for-quick-sale and special offers on breads. Using different breads adds variety to sandwiches and makes them more interesting than just sticking to your sliced white. Try multi-grain, seeded breads, big up bagels, appreciate pita and wonder at wraps – the variety is endless.

Don't forget about leftovers too. Some starchy foods are great hot or cold: leftover pizza, pasta salad or couscous and rice dishes can all work well as lunchtime pack-ups.

Packing in some protein

Go for a variety of proteins including lean meats, chicken or turkey, fish, eggs, nuts and beans and pulses.

Experiment with combinations:

✔ Cottage cheese and raisins

✔ Peanut butter and banana

✔ Roast beef, watercress and horseradish

✔ Tuna and beetroot

✔ Turkey, pesto and tomato

Getting fruity

Packed lunches make it really easy to boost your fruit intake so you can get your five-a-day of fruit and veg. And fruit often comes in its own wrapper – what could be simpler than that!

Try these ideas to help add a touch of fruitful nutrition to your lunch box:

- ✔ Keep it as nature intended – a simple apple, pear or banana is portable and makes a quick and easy lunchtime dessert.

- ✔ Don't forget that dried fruit counts towards your five-a-day. A tablespoon of dried fruit is one portion, so pack some raisins, soft dried apricots, dates or prunes to pick on at your PC.

- ✔ Crudités are great in lunch boxes. Dip carrot sticks, celery, cherry tomatoes and grapes in low fat hummus or salsa or just enjoy them on their own.

- ✔ Canned fruit often come in mini cans with ring pulls – perfect for a lunchtime fruit cocktail.

Diving into dairy

Bones and teeth need around three servings of dairy foods a day to get enough calcium. A serving includes:

- ✔ Individual cheese portion, such as mini Edam

- ✔ Low fat fromage frais

- ✔ Low fat fruit or plain yoghurt

- ✔ Low fat milkshake

- ✔ Small pot of rice pudding or low fat mousse

Sweetening the deal

There's nothing wrong with having something sweet to finish lunch. Here are some delicious, healthy sweet treats that won't break the calories bank:

- ✔ Cereal bar

- ✔ Fruit bread, scones, currant buns or fruit-filled rye crackers

- ✔ Fun-size bar of chocolate

- ✔ Plain biscuits like a digestive or rich tea

Dropping in a drink

Choose from:

- ✔ Dilute, no-added-sugar squash
- ✔ Fruit juice
- ✔ Plain water – still or sparkling
- ✔ Skimmed or semi-skimmed milk
- ✔ Smoothies
- ✔ Soup in the winter
- ✔ Tea or coffee

Considering what's in your morning coffee

When did stopping for a mid morning cup of coffee become more confusing than choosing from a fancy French restaurant menu? The simple cup of coffee has had a major makeover in the past few years. Coffee chains are on every corner, all touting their white chocolate macchiato, double chocolate mochas or cinnamon and honey frappacino. Did you know, though, that some of these sophistacoffees can add as much as 600 calories to your daily intake? That's 30 per cent of a women's guideline daily amount (GDA) or the equivalent of the calories she's recommended to eat at lunch or dinner.

Meeting a mate for a natter over your favourite brew once in a while is, of course, not a problem, but picking up a frothy, cream-topped, extra syrup caffeine fix every morning on the way to the office could be a recipe for a spare tyre.

First, here's the lowdown on what all the fancy names mean:

- ✔ **Espresso:** Hot water forced through ground coffee to give a strong, almost calorie-free caffeine hit
- ✔ **Americano:** Espresso shot with enough hot water to fill the cup
- ✔ **Macchiato:** Espresso shot with a small amount of frothed milk
- ✔ **Latte:** Espresso shot with a lot of hot milk to fill the cup
- ✔ **Cappuccino:** Espresso shot with a lot of milk foam to fill the cup – sometimes topped with cinnamon or cocoa

Now, none of these coffees are too much of a problem as they stand, especially if you ask for skimmed milk to make your latte or cappuccino (usually called a *skinny* latte or cappuccino). It's the sugar rich syrups, flavours and whipped cream that make the difference and drive up the fat, sugar and calorie totals.

So find the coffee you like but make it skinny most of the time if it's with milk, and keep the extras for special occasions and treats.

Keeping it fresh

Forget soggy sarnies with melting marg. Here are some tips for a cool lunch:

✔ Use a cool bag and put in an ice pack.

✔ Freeze a carton of juice and place it in with your food till lunchtime.

✔ Keep any food you prepare the night before for work in the fridge overnight.

✔ Don't store your lunch near a radiator or on a sunny window ledge.

Watching what you drink

It's not just what you eat and drink during the working day that's important when it comes to being healthy: you've also got to consider the after-work wind-down with colleagues in the pub. Now, we like a glass of wine along with everyone else and we don't want to be killjoys, but every now and then it's a good idea to remind yourself how much alcohol is too much.

Experts recommend a maximum of two to three units of alcohol a day for women and three to four units a day for men. Oh, and that advice doesn't mean it's safe to go booze free all week and then blow all your units in one night – binge drinking can be very bad news for the body!

More and more drinks show the number of units they contain in a typical serving these days, but for the record one unit contains 8 grams or 10 millilitres of alcohol. The percentage volume of alcohol on the label tells you the number of units in 1 litre. So a 12 per cent wine contains 12 units in a litre or 9 units in a standard 750-millilitre bottle, or 1.5 units per 125-millilitre glass.

Very few pubs serve wine in 125-millilitre glasses these days. Some large glasses of wine served in a pub are the equivalent of a third of a bottle!

If you enjoy the social aspect of the after-work drink but don't want to overindulge, try the following tips to limit your intake:

✔ Alternate an alcoholic drink with a non alcoholic one.

✔ Make sure you drive to work or the station so you can't be pressured to drink.

✔ Sip your drink slowly to make it last longer.

✔ Order non-alcoholic beer or wine.

✔ Don't top up your glass before you finish a drink so you keep an eye on how many units you've had.

✔ Try white wine spritzers mixed with soda or sparkling water.

We talk more about alcohol in Chapter 8.

Fuelling Up: Nutrition on the Road

Surveys suggest that time spent on the road has increased dramatically over the past ten years, with each person in the UK spending an amazing 225 hours per year travelling by car.

Thanks to the warning signs people are aware that 'Tiredness can kill' and know to take a break at least every couple of hours to rest and recharge. Pit stops are a great opportunity to refresh both your body and mind, but either the car becomes a second kitchen/dining room or you have to purchase what you need *en route*. Neither option is ideal if you're trying to eat healthily, but with a bit of planning you can limit the damage and fill that gap with healthy choices on the road.

Taking food with you

Packing a healthy picnic is a great way to stay healthy and saves you money. Many main roads have laybys and picnic areas where you can stop, and all motorway services provide an outdoor area where you can eat your own food. Even if you eat in the car, do make sure you at least get out to stretch your legs and get some fresh air when you stop. Check out the RAC website (www.picnicwithrac.co.uk) for some advice on how to find a good spot for a picnic.

Here are some tips for an enjoyable, healthy picnic:

- ✔ Keep it simple: sandwiches and salads are ideal.
- ✔ Take a cooler bag to keep items fresh.
- ✔ Don't forget the napkins or wetwipes to make clearing up easier.
- ✔ Take your own tea, coffee or even soup in a flask.

And in case you're hungry while on the move, pack some healthy snacks that can keep you going like wholegrain crackers or oatcakes, dried fruit or fresh fruit. Raw vegetables sticks (try celery, carrot cucumber or peppers as crudités) are great chopped up small so they're easier to eat.

Selecting from service stations

When you're running on empty along the motorway you sometimes have no choice but to grab a bite at a service station to refuel both you and the car. The first service stations opened over half a century ago and were quite glamorous affairs. Watford Gap was the first 24-hour restaurant in Britain and famous bands, including the Beatles and the Rolling Stones, went out of their way to visit in the '60s and '70s.

Unfortunately, service stations have since become synonymous with greasy, fast food with the fancy coffee costing more than the petrol. But recently, thanks to increased customer demand, it seems standards are improving as retailers become aware that consumers want more choice alongside the chocolate, pasties, chips and burgers. Many service stations now offer healthy options like wholegrain toast and cereal, porridge, fresh fruit, omelettes and even smoked salmon and scrambled eggs alongside the full English breakfast. Others feature salad bars, jacket potatoes, homemade soup and pasta as an alterative to the fish and chip lunch. And some have even been revamped by celebrity chefs and have an entry in the *Good Food Guide*!

Nevertheless, the food in service stations often comes at a price. Many of the main course meals are expensive and a recent Consumer's Association survey found bottled water for sale at four times the price in a local supermarket and cappuccino at £3 a cup! But cheaper options exist, if you look for them. It's worth knowing that AA and RAC members are entitled to good discounts on food and drink bought at some service stations if they show a membership card. Check your member's handbook for details. Another useful source of information is `www.5minutesaway.co.uk`. This site is regularly updated by frequent motorway travellers, and it highlights cheaper places to eat and drink that are less than five minutes' drive from motorway junctions.

In addition, a welcome trend in recent years has been the opening of small branches of supermarket outlets in service stations. Customers can access healthy choices for items such as sandwiches, snacks and drinks, or find fruit and vegetables at a reasonable price. These are often ready to eat in individual portions with plastic cutlery and paper napkins, making it easy to picnic in the outdoor areas when it's fine. And if you fancy a hot drink, remember that often service stations sell these in vending machines at more reasonable prices than the coffee bar.

Here are some suggestions for healthy choices in service stations:

- ✔ Bottled water
- ✔ Cereal bars
- ✔ Fresh fruit including fruit salad
- ✔ Fruit juice and smoothies
- ✔ Individual cereal pots with milk or yoghurt
- ✔ Low calorie hot chocolate
- ✔ Low fat milk
- ✔ Low fat savoury snacks like plain popcorn, pretzels and trail mix

> ✔ Low sugar squash and fizzy drinks
>
> ✔ Pre-packed healthy eating sandwiches or wraps
>
> ✔ Ready-to-eat salads with dressing on the side
>
> ✔ Tea or herbal/fruit teas

Many of these items have colour-coded pack labels to show various nutrients such as fat, saturated fat, sugar and salt. Red means high, amber is moderate and green is low (traffic lights aren't just confined to the road these days!). Pick those with mostly green and amber symbols if you want to find healthy choices at a glance.

If you want to plan in advance to stop at a service station supermarket you can find details of locations at the Motorway Services website (www.motor wayservices.info).

Staying alert

Dehydration is a real risk in a hot car, even one with air conditioning. Even slight dehydration can affect your concentration, alertness, decision-making ability and reaction time, all of which are vital while driving. So top up your fluid levels regularly. Don't rely on thirst – drink before you feel thirsty and keep a water bottle handy on long journeys, especially in hot weather.

The best drinks to choose are water, smoothies, fruit juice, low sugar squash or diet fizzy drinks. Caffeinated drinks help some people concentrate, and contrary to popular belief they do count towards fluid intake. The Department for Transport THINK Road Safety advice encourages drivers to have two cups of coffee or another caffeinated drink alongside a short nap (no more than 15 minutes), to allow time for the caffeine to kick in, before continuing their journeys. Regular drinkers of tea, coffee, cola or energy drinks don't suffer from the mild diuretic effect of caffeine. However, loading up on caffeine doesn't mean it's safe to drive harder or for longer.

What you have to eat before you set out can also have a positive effect on your brain performance when going on a long journey. For example, research has shown that you can significantly increase your ability to concentrate throughout the morning by eating a low glycaemic index (GI) breakfast (see Chapter 7). Examples of good low GI foods to start the day with are wholegrain breakfast cereal, porridge, muesli with milk, yoghurt or fresh fruit, or wholegrain or seeded breads for toast.

Feasting on fast food

Fast food restaurants don't conjure up the best healthy eating credentials in most people's minds. Generally, fast food meals are high in calories, salt and fat. They're often criticised for lacking vitamins and minerals, and until recently chips were the only vegetables on some fast food restaurant menus. The good news is that many of these food outlets are now offering healthier menu options.

Everyone eats fast food from time to time, so it's very important to know your way around a menu to limit the damage. Knowing what types of menu items are healthier than others is a good start. Here are some tips for eating out at a fast food outlet:

- ✔ Order food to go. Studies reveal that people eat less at home, so take your pizza home and have a piece of fruit afterwards.

- ✔ Avoid buffets. All-you-can-eat buffets promote exactly that.

- ✔ Don't be afraid to special order. Speak up for yourself – most restaurants certainly won't force you to have a salad dressing with your salad.

- ✔ Watch portion sizes. Super sizing has become the norm, and a regular portion can contain double the recommended calories, fat and salt you should be eating at one sitting.

- ✔ Sharing starters and desserts is a great idea. It'll satisfy you and save you calories.

- ✔ Don't forget the calories in drinks. A milkshake could easily contain 13 teaspoons of sugar.

- ✔ Don't add extra salt. Most of these foods contain a lot of salt added during processing. Healthier fast food options do exist. Look out for:
    - Baked potato (with vegetables)
    - Fruit or fruit salads
    - Grilled chicken or fish sandwich
    - Salad with dressing on the side or fat free salad dressing
    - Single hamburger
    - Whole-wheat rolls
    - Wraps on whole-wheat tortillas (without dressing)

Table for two?

An incredible choice of places to eat exists these days. In the past, eating out was a rare treat and licence to have a blow out, but today people eat out far more than ever before. So if you want to stay healthy, you need to be a little more savvy these days when you peruse a menu.

Here are some top tips to make sure you stay on the right side of the tracks in restaurants:

✔ If you're going out for dinner make sure you eat properly throughout the day and don't skip meals. If you're really hungry when you sit down to dinner it'll be hard not to resist rich, indulgent foods.

✔ If the restaurant has a website check out the menu before you go out to see what looks like a good choice.

✔ Ask for some water when you sit down at the table – a drink can often take the edge off hunger.

✔ Have a piece of bread or some breadsticks while you look through the menu.

✔ Instead of having three courses, how about a starter and main or a main and dessert?

✔ Ask for sauces and dressings on the side so you control how much you have.

With so much choice around it's a good idea to have an idea of the healthier options in different styles of restaurants. Now this could almost be a whole Dummies book of its own, but here we stick to just a few tips for some of the most popular cuisines.

Idyllic Italian

Pizza and pasta is a surprisingly good choice – the starchy carbohydrate is filling but it's the toppings and sauces to watch out for.

✔ Choose tomato sauces for your pasta.

✔ Top pizzas with veggies, fish, lean meats or chicken and avoid the four cheese, salami or pepperoni toppings.

✔ Go for thin and crispy or deep crust bases, but watch out for stuffed crust.

✔ Order a salad on the side – mixed, green or tricolor with mozzarella, tomato and avocado are all great.

Make mine a Mexican

Lots of chilli, spice and sombreros, but do you know your quesadillas from your chimichangas? Many Mexican dishes involve corn or flour tortillas either as flat stuffed delicacies like quesadilla or as stuffed parcels like chimichangas and burritos. The stuffings are usually spiced chicken, beef or beans, and they're often topped with guacamole (avocado), sour cream and cheese.

✔ Go for burritos, enchiladas and quesadilla more often than chimichangas, which are fried while the others are baked, and go easy on the toppings – ask for them on the side.

✔ Salsa is all veggie and low in fat, so use this as a topping more than the creamy, high fat toppings.

Calories on the menu

In 2008 the Food Standards Agency surveyed 2,000 consumers and over 85 per cent of them said they wanted restaurants, pubs and cafes to show what's in the food they serve and that this information should be available in the menu or at the point-of-sale point. So in 2009, 21 familiar high-street food service chains introduced calories on the menus.

Researchers looked at the calorie-informing scheme from the consumers' and the retailers' points of view, and found mixed responses:

✔ Consumers still thought of calories as being connected with dieting and weight loss, so struggled to see that the calorie information was relevant unless they were trying to lose weight. Consumers also lacked understanding of the recommended daily allowance of calories and how much exercise you need to burn off calories.

✔ Overall, retailers saw the initiative as a positive thing, but from a business perspective there were some issues. To take part in the scheme retailers needed to pay for sophisticated IT systems to nutritionally analyse recipes and unique ingredients, and the information-gathering process was time consuming. And any changes to the information displayed needed to coincide with production of marketing and packaging materials, which are sometimes produced six months ahead of use.

The FSA continue to evaluate the scheme at each stage of its development. More retailers are joining the scheme, so it's likely that it will become a more common feature of menus and point-of-sale material. However, it seems that communication to consumers must improve if the calorie information is to have any significant impact on the food choices people make when they're out and about.

Choosing Chinese

Chinese food, can be healthy – fast cooking at high heat means food is in contact with fat for a very short time and retains a lot of its vitamins and minerals. However, some Chinese food is loaded with far more salt and sugar than you might realise.

✔ Stick with clear soup starters and watch out for prawn crackers, sesame toast and spring rolls, which are all deep fried.

✔ Black and yellow bean sauces are good main courses as are chow mein dishes and szechuan. Watch out for battered and sweet and sour dishes, which hide a lot of fat and sugar.

✔ Choose plain boiled or jasmine rice and avoid the high fat fried rice as a side dish.

Indulging in Indian food

Even more popular than the traditional Sunday roast, Indian cuisine leads the way for UK taste buds. Lots of healthy choices exist, as do a few pitfalls to watch out for.

✔ Tikka, bhuna, tandoori and tomato-based curries with chicken, fish or prawns are all good choices. Watch for tikka masala and korma dishes, which are creamier and higher in fat, and some lamb dishes can be a little on the fatty side too.

✔ Go for plain naan bread or chapatti rather than filled naans.

✔ Choose Bombay potatoes or saag aloo side dishes and give the fried onion bhaji a miss more often.

✔ Plain rice over pilau rice is a good choice.

Although everyone does it more often these days so it's not such a big deal, eating out is still meant to be a pleasure. Don't be too rigid about what you choose, and every now and then go for it and have exactly what you fancy. We've just given you suggestions for when you do want to be a little more choosy, but they're not rules!

Chapter 17

Looking at Standards in Schools and Other Institutions

In this Chapter

▶ Ensuring healthy meals, snacks and drinks for youngsters in nurseries and schools

▶ Following dietary guidelines for people in hospital and residential care homes

▶ Exploring nutritional standards for the all-day provision of food for those in the armed services and in prisons

*I*n the UK food paid for from the public purse accounts for over a billion meals each year. However, nutrition surveys reveal that the diet of many of the groups receiving these meals is poor, with intakes of saturated fat, sugar and salt all way too high and some vitamin and mineral intakes too low.

The Department of Health estimates the costs of treating the effects of poor diet to be around £4 billion per year. In an effort improve public eating habits, agencies such as the UK Food Standards Agency and the School Food Trust have in recent times explored ways to improve meals provided in public institutions.

The four biggest public sector food service areas in the UK are nurseries and schools, the NHS, the Prison Service and the Armed Forces. Quality of food served in these settings affects not only service users, but also those who work in or visit them. As a result of this work, some useful standards and guidance for food provision are emerging. These vary according to the specific population served and whether they cover all-day provision of food or individual eating occasions such as the school lunch. The standards may be food based and in line with the Eatwell plate (see Chapter 14), such as providing at least two portions of fruit and vegetables per meal or having fish on the menu twice a week. The standards may also be *nutrient based* (defined in relation to the relevant dietary reference values or DRVs; see Chapter 15), such as requiring that a meal must provide at least 35 per cent of the reference nutrient intake (RNI; see Chapter 15) for calcium or no more than 11 per cent of food energy from sugar or saturated fat.

Some institutions write food and nutrition standards into catering contracts and monitor them. A wealth of guidance is available to help caterers achieve standards – such as on the types of food from each group to include, the procurement of healthier manufactured foods and ingredients for use on the menu in, and suggestions for improved cooking and serving practices including suitable portion sizes.

After healthier food choices start to become more available the need still exists to ensure the food is attractive and tasty, and to encourage uptake by the service users. Strategies may include promoting specific choices as the healthy dish of the day, using traffic lights or calorie levels to highlight the nutritional value of meals on the menu, or offering incentives like reward points for including healthy choices such as a portion of fruit and vegetables.

For many service users in institutions, food is a highlight of the day, and in addition to health it's important to consider choice, food preferences and cultural or religious needs. Issues such as sustainability, local procurement and cost are also increasingly on the agenda in menu planning in many institutions. Set against this background, this chapter explores some of the major food issues for a range of public institutions, highlighting some of the key nutritional areas relevant to each.

Knowing about Nutrition for Tiny Tummies

How you eat in the first years of life has a lasting impact on how you grow, and on your long-term health as you move into adulthood. Taste preferences and dietary habits formed in the first few years last a lifetime.

The period from 1 to 5 years sees more growth and development than at any other stage of human life. This requires energy, nutrients like protein, and vitamins and minerals. But during these early years, a lot can go wrong. Children have small appetites and learn to eat by the example set by others around them. They may go through periods when they go off their food because they're unwell or trying to assert their authority. Face it, when it comes to what's for dinner, children under 5 are usually reliant on adults making the right choices for them.

Yet those providing food to children don't always get it right. Nutritional disorders in the under 5s are on the increase. Iron deficiency anaemia, rickets, growth problems, tooth decay, constipation and childhood obesity are becoming more and more common.

Establishing a healthy diet from the start can reduce the risk of heart disease, diabetes and some cancers in adulthood. And yet institutional nutritional standards overlook this critical window of nutritional opportunity. More children than ever in the UK attend nursery education, and with over 730,000 nursery places available, many children attend nursery five days a week and consume more of their meals in childcare than in their own home. But the nutritional quality of the food children eat at nurseries is a total lottery.

Realising the state of children's diets

Nutritional problems are common in the under 5s, with many deficiency diseases on the increase at alarming rates including some diseases that had almost been eradicated a few decades ago.The National Diet and Nutrition Survey and the more recent Avon Longitudinal Study of Pregnancy and Childhood highlight the direct links between what you feed a child and the short-and long-term health consequences. They found that in under 5s:

- ✔ Over 80 per cent of have a low iron intake.
- ✔ Over 70 per cent have a low zinc intake.
- ✔ Fifty per cent have a low vitamin A intake.
- ✔ Eighty-five per cent eat too much sugar.
- ✔ Those from low income families don't eat enough vitamin C or folate.
- ✔ Only 20 per cent take a daily vitamin D supplement, despite government guidance that all under 5s should do so.

They also found that almost 20 per cent of 2- to 5-year-olds are obese, and another 14 per cent are overweight and on their way to becoming obese. If this trend continues, by 2050 it's estimated that 25 per cent of children in the UK will be clinically obese.

Clearly young children are a vulnerable group and it's essential for them to receive adequate nutrition for health and growth. So with more and more children entering childcare, it makes sense to set standards for the food children eat while at nursery.

Noting the lack of nutritional standards for young children

In the UK, only Scotland has comprehensive national guidelines for 1- to 5-year-olds in childcare, but these aren't mandatory. Some organisations have produced policies and guidelines, but as yet nurseries don't have to follow these and no external body monitors nutrition in childcare. The bottom line is that it's up to nursery managers what gets served to children, but often they have no nutrition qualification or training, and if they look for guidance it's vague and incomplete.

The government and expert organisations have gone some way to recognising that some sort of nutrition is important for this age group, but not far enough. For example:

- Healthy Weight, Healthy Lives, a government strategy that focuses on the health of the population, talks about the importance of breast-feeding, weaning and healthy toddler meals, but gives no guidance on what that a healthy diet actually means.

- Two government strategies – the Child Health Promotion programme and Healthy Lives, Brighter Futures – emphasise the importance of a healthy diet for the under 5s but go no further in saying what that diet should contain.

- The National Institute for Clinical Excellence (NICE) sets the standards of treatment to tackle various health issues and has made recommendations on child nutrition and obesity, but only for 2-year-olds. This misses the weaning period when children move from milk to solid food as the main source of nutrition – the critical time to get good dietary habits started.

- The Food Standards Agency advice is vague: they talk about the food groups but don't say how much of each food group under 5s should be eating (but they don't generally give this information for other age groups either).

- The Department of Health Change 4 Life programme (http://www.nhs.uk/change4life/Pages/Default.aspx) has some useful practical advice on small, easy steps you can make to help set your family on the way to a healthy diet.

On the positive side, in 2006 the Caroline Walker Trust, a voluntary organisation, issued very specific, clear, nutrient-based standards for young children. Dieticians and health professionals see *Eating Well for Under 5s in Childcare* as the gold standard in nutrition for this age group. The standards have been adopted by organisations like Grub4Life (www.grub4life.org.uk), which develops training, menus and recipes for hundreds of nurseries and children's centres throughout the UK.

In addition, computerised nutritional analysis programs are available to help cooks and nursery managers put together balanced menus for children, but the user needs a certain amount of training to use the program effectively.

Using a food-group approach

Some people feel that nutrient-based standards are too complicated and detailed for the under 5s and that simpler food group guidance is more appropriate. We agree. So in Table 17-1 we give you a simple food-group approach to healthy eating in children that you can use as guide when assessing your child's eating, particularly at nursery.

Table 17-1	The Food Groups for Under 5s Menu Planning		
Food Group	*Foods Included*	*Main Nutrients Supplied*	*Recommendations for 1- to 4-year-olds*
Bread, cereals and potatoes	Bread, chapatti, breakfast cereal, rice, couscous, pasta, millet, potatoes, yam, foods made with flour such as pizza bases, buns, scones, pancakes, crackers	Carbohydrate, B Vitamins, fibre, some iron, zinc, calcium	Offer these at each meal and some snacks.
Fruit and vegetables	Fresh, frozen, tinned and dried fruits and vegetables, pure fruit juices	Vitamin C, other antioxidants, fibre	Offer fruit at breakfast and fruit and vegetables at each main meal. You can also either at snack times. 1 portion = what fit's in the child's hand
Dairy foods	Milk, yoghurts, cheese, fromage frais, calcium enriched soya milks, tofu	Calcium, protein, iodine, riboflavin	Offer three servings a day. One serving is 120 millilitres of milk in a beaker or cup, one pot of yoghurt or fromage frais, a serving of cheese in a sandwich or on a pizza, a milk-based pudding or a serving of tofu.
Meat, fish and alternatives	Meat, fish, eggs, pulses, dhal, nuts, seeds	Iron, protein, zinc, magnesium, B vitamins, vitamin A, omega 3 fats in oily fish	Main meals (apart from breakfast) should always contain an item from this group. Offer fish twice a week and oily fish at least once a week.
Foods high in fat/sugar	Cream, butter, margarines, cooking and salad oils, mayonnaise, cakes, biscuits, chocolate, honey, jam	Some foods provide vitamins D and E and omega 3 fatty acids	You can offer small amounts at meals in addition to but not instead of the other food groups.
Fluid	Drinks	Water, fluoride in areas with fluoridated tap water	Offer with each meal and snack. Offer extra in hot weather or after physical activity.
Vitamin supplements			Offer a supplement for up to age 5 containing vitamins A and D.

Finding a nursery that knows nutrition

If you're a parent with a little one who's about to start nursery, we're not suggesting you burst through the doors like the nutrition police to see what they serve up for lunch! But you can look for a few simple things to see how seriously the nursery takes nutrition, and what they do to make sure they're getting it right.

When considering a prospective nursery, do the following:

Ask if they have a nutrition or food policy:

- ✔ **Look at the menu display.** Does the nursery display menus so that parents know what's being served each day? Not only does this help you see whether they follow the food group guidance; it also helps you plan the food you serve at home. And hey, you might also get some inspiration for meals to make at home.

- ✔ **Check the website.** Most nurseries have a website these days and those who are proud of the food they serve want to tell prospective parents (customers) about their good work. If they don't have a website, have they got something to hide?

- ✔ **Ask which standards the nursery uses.** If you go along to meet the manager or attend an open day at the nursery, be brave and ask questions. You would about issues like education or discipline, so why not ask about food? Here are some questions you can ask:

  - Who writes the menu?

  - Do you work to any specific nutrition standards?

  - Do you have an in-house nutrition policy?

  - Do you work with a dietitian?

- ✔ **Find out about the healthy eating award.** A lot of local authorities and corporate nursery groups have Healthy Eating Award schemes. The criteria for the awards varies from location to location, but at least if a nursery has an award it gives you an indication that someone in the organisation is thinking about nutrition. Find out whether a scheme operates in your area and ask whether your nursery has received an award – and if not, why not?

Setting the Standard in Schools

Schoolchildren of all ages are growing and developing rapidly and so need a range of nutrients. This is a key time for formation of dietary habits and those adopted during infancy and preschool start to become ingrained, often lasting into adulthood. Unfortunately, as with preschool children, dietary surveys on this age group show intakes to be particularly poor. Intakes for several key nutrients such as iron, calcium, folate and vitamin A are low alongside relatively high intakes of saturated fat, salt and sugar, and poor intakes of fibre, fruit and vegetables. Diet quality is a particular concern in adolescents, many of whom are also physically inactive and adopt other health risk behaviours such as smoking and drinking alcohol.

Schools are an excellent setting in which to favourably influence a range of health-promoting behaviours including diet. Opportunities exist with the school lunch, with guidance for packed lunches and with food served at other occasions such as breakfast and after school clubs, at breaks and from tuck shops and vending machines. Schools can also educate on the importance of healthy eating and on food preparation skills via the curriculum. In recent years excellent work across the UK has looked at a whole school approach in all these areas.

Considering food served in schools: Lunches and snacks

School meal standards were first introduced in the 1944 Education Act to provide a third of daily nutritional requirements. Use of these standards continued until 1980, when they were abolished. But in 2006 the government reintroduced nutrition based standards that cover all food and drink served in UK schools – both local authority and independent. The standards have been phased in gradually over several years in both primary and secondary schools. The aim is to improve lunches by:

✔ Using better quality and more nutritious ingredients

✔ Providing healthier and more balanced meals

✔ Replacing less healthy food choices, high in fat, salt and sugar with more nutritious options

Offering healthier foods with increased vitamin and mineral content

Table 17-2 shows the government's food-based standards, which relate to foods that schools can and can't serve.

Table 17-2	Food-based Standards for Schools
Foods	*Standards*
Fruit and vegetables	No less than two portions offered per meal, to include at least one portion of salad or vegetables and one portion of fruit
Red meat	On the menu twice weekly in primary, three times a week in secondary
Fish	On the menu weekly in primary, twice weekly in secondary
Oily fish	On the menu at least once every three weeks
Deep-fried foods	No more than two deep-fried items offered in a week; if fish and chips is provided then no additional fried foods allowed in the same week
Processed meat products	Limited on menu and minimum meat content specified
Bread	Freely available at lunch meal without spread; other starchy foods not cooked in oil available at every meal
Confectionery and savoury snacks	Not available at lunch; should be low fat, low sugar
Salt	Not available at lunch table or service counter
Drinks	Only water, low fat milk, pure fruit juice or smoothies (no full fat milk or fizzy beverages)
Water	Easy access to free chilled drinking water throughout day

Source: Adapted from the School Food Trust standards information (www.schoolfoodtrust. org.uk)

Schools must also meet the nutrient-based standards (Table 17-3), which relate to maximum levels for fat, saturated fat, sugars and salt and minimum levels for other key nutrients such as fibre, vitamin C, iron and calcium.

Table 17-3	Nutrient-based Standards for School Lunches in Primary and Secondary Schools		
Nutrient	*Value*	*Primary Schools*	*Secondary Schools*
Energy (kcals)	50% EAR	530	646
Fat (g)	< 35% food energy	20.6	25.1
Saturated fat (g)	< 11% food energy	6.5	7.9
Salt (g)	< 30%	1.25	1.8
Added Sugar (g)	< 11% food energy	15.5	18.9
Fibre (g)	> 30%	4.2	5.2
Protein (g)	>30% RNI	7.5	13.3
Iron (mg)	> 35% RNI	3	5.2
Zinc (mg)	> 35% RNI	2.5	3.3
Calcium (mg)	> 35% RNI	193	350
Vitamin A (mcg)	> 35% RNI	175	245
Vitamin C (mg)	> 35% RNI	10.5	14
Folate (mcg)	> 35% RNI	53	70

EAR: Estimated Average Requirement; RNI: Reference Nutrient Intake.

*Source: Adapted from the School Food Trust standards information (*www.schoolfood
trust.org.uk*)*

And research exploring the effects of school meal standards indicates these standards are benefiting kids. One study on secondary schools carried out by the School Food Trust used independent observers to examine the levels of concentration and engagement shown by pupils in various tasks. They found pupils in schools implementing the standards were significantly more likely to be on task with work compared with those in schools not yet adopting the standards.

Thinking about food brought from home: Packed lunches

On average around half of school pupils in the UK take a packed lunch from home. If parents choose to give their children packed lunches instead of school meals it can be a challenge not just to keep lunches varied but also to ensure they meet similar standards to school lunches.

Compared with school meals, packed lunches are less likely to contain the recommended nutrient intakes for protein, fibre and key vitamins and minerals. In addition, when members of the British Dietetic Association looked at the lunchboxes of over 550 children, they found that levels of saturated fat, salt and sugar were much higher than advised, coming in part from crisps and chocolate biscuits or cake. The average lunchbox contained the equivalent of eight teaspoons of sugar and fewer than half contained a piece of fruit.

As a result many schools have, after consultation with pupils and parents, adopted packed lunch policies to improve the standards of food brought from home. Many of these policies contain guidance on how to select healthier items for the lunchbox. The school in turn agrees to ensure pleasant surroundings in which to eat packed lunches and free fresh drinking water. Some schools review the contents of lunchboxes regularly and reward healthier choices with stickers or certificates for younger pupils.

Here are some tips for healthy packed lunches:

✔ Each day include:

- At least one portion of fruit and one of vegetables, such as salad, homemade or salt-reduced vegetable soup, homemade pizza with vegetable toppings or fruit salad

- Lean protein sources such as lean meat (ham, turkey or chicken), fish, pulses or nuts

- A starchy food such as bread (wholegrain, half and half, pitta, bagels), pasta, rice, couscous or potatoes

- A portion of lower fat dairy like milk, cheese, yoghurt or fromage frais, low fat custard or rice pudding

- A healthy, low sugar drink. Examples include water, fruit juice, low fat milk, yoghurt drinks or smoothies

✔ Once every three weeks include oily fish

✔ Avoid including crisps, confectionery or chocolate-coated biscuits. Try instead nuts, seeds, dried fruits (raisins, figs, prunes or apricots), vegetable crudités, crackers, breadsticks, scones, currant buns or malt loaf. Include sausage rolls, pies and pasties only occasionally

Serving Up Better Hospital Food: Leading by Example in the NHS

Food plays an important role in the lives of many hospital inpatients and can be an important part of their treatment. Poorly nourished patients take longer to recover from trauma, surgery and illness, they stay in hospital longer, they develop more complications and they're more susceptible to hospital superbug infections. Surveys suggest that large numbers of those admitted to hospital are already malnourished, and as many 60 per cent are at risk of their situation worsening during their stay. Good nutrition can not only reduce these problems, but can shorten the length of hospital stays, thus saving the NHS money.

The amount spent per day on providing food for hospital patients varies around the UK. However, the average figure is £2.60 per head per day for all meals and drinks. This goes some way to explaining the difficulties inherent in meeting the challenges of better hospital food.

And malnutrition isn't the only consideration for caterers in hospitals. Many NHS settings also serve the needs of patients for whom the emphasis may be more on healthy eating. This group might include long-stay patients in psychiatric units, maternity or paediatric inpatients, or staff who eat most of their meals and snacks on the premises.

In addition, a strong feeling exists among the public and NHS staff that poor quality nutritional care in hospital is an affront to dignity and must not be tolerated.

In response to these varying demands, in recent years government advisors have closely monitored the provision of hospital food, launched a multitude of initiatives and made many recommendations. NHS catering systems are starting to respond to a broad range of legislation and guidance.

The school curriculum

In a recent Food for Thought survey researchers found that 99 per cent of adolescents knew how to use a DVD player and 82 per cent knew how to use a microwave, but only 58 per cent knew use a vegetable peeler and just 43 per cent could boil an egg! As a result of shortfalls in skills and knowledge around healthy cooking and eating, many schools have started to reintroduce practical healthy-cooking skills to the curriculum with the aim of increasing food independence in young adults so they're less reliant on takeaways and ready meals. Many agencies, including the UK Food Standards Agency, have produced useful guidelines for the competencies young people need to help them make healthy food choices. These tend to be age- and theme-specific for teachers' use, but general themes include:

- Appreciating the need for variety and balance in diet

- Knowing that you need to eat some food groups in greater proportion than others

- Recognising that foods provide different nutrients, some of which you need more at particular life stages

- Using dietary information to chose foods

- Being aware that food labels can aid healthy choices

Department of Health standards for NHS catering include a commitment to 'provide choice and a safely prepared and balanced diet'. Their 1995 Health of the Nation nutritional targets for hospital catering produced laid down useful guidance on minimum levels for energy and protein both in individual meals and across the whole menu. However, these standards didn't include a broader range of nutrients. But recent initiatives been more positive, such as the Better Hospital Food Programme that introduced innovative and tasty 'chef's dishes' on menus, a delivery of three meals and two snacks per day and 24-hour access to food for those missing a meal via snack boxes or light meals.

Even more recently, in a landmark report the Council of Europe defined and highlighted the key characteristics for how good nutritional care should look in hospital settings. Recommendations include:

- The screening of all patients on admission for malnutrition
- Drawing up a nutritional care plan that states how the hospital will meet patients' nutritional needs during their stay
- Flexible and patient-centred catering facilities that offer 24-hour care
- Staff training on the importance of nutrition and how to meet varying needs of their patients
- A policy of protected mealtimes, whereby patients are given enough time and an environment conducive to enjoying their food without interruptions for routine medical care

In line with such recommendations, hospital food seems to be improving steadily and many commentators believe that it's better now than at any time since monitoring began. However, evidence of poor practice in some NHS settings still exists. In a recent survey of inpatients only 58 per cent of those who needed assistance to eat their meal said they always received it. Evidence suggests that protected mealtimes aren't implemented in all NHS settings, and that medical staff don't always adhere to the policy even if it's in place.

Visiting friends or family in hospital – how can you help?

If you are visiting family or friends in hospital there are several things you can do to help. Remind them of the importance of food and drink in aiding their recovery and encourage them wherever possible . You could for example help them to take extra drinks or snacks between meals or fill in their menu card with them looking for the most nutritious choices. Bringing in a favourite snack item may be helpful and even if it doesn't seem like a 'healthy choice' it will help to boost their energy intake. However, do ask ward staff for guidance before you bring in any food or drink as perishable food may not be permitted if there are limited storage facilities to keep it fresh. Food that can be safely brought in for patients include:

✔ Shop-bought cakes and biscuits or savoury crackers in sealed packets with clear use by dates

✔ Pre-washed or easy peel fruit – choose types that are easy to eat

✔ Fruit squash, canned drinks, individual serving cartons of long life fruit juice and bottled water

✔ Dried fruit and nuts

✔ Sweets or chocolate

The Food Standards Agency has useful guidelines on food- and nutrient-based standards for children, adults and older people in hospital at `http://www.food.gov.uk/multimedia/pdfs/walkertrustreport.pdf`.

Caring for Older People: Standards in Residential Homes

Over 400,000 older adults live in residential care homes across the UK, including one in four of all adults over the age of 85. Studies have highlighted that often these residents are among the oldest and frailest in society. Food provision is of key importance because many are already malnourished prior to admission, having been subject to risk factors in the community such as illness, swallowing problems, multiple medication use, poverty, social isolation or psychological problems including dementia. Dietary surveys of care home residents highlight particular concern over intakes of folate, riboflavin, vitamins C and D, iron, zinc and potassium from diets that are high in salt and low in fruit, vegetables and fibre.

A recent survey showed that older adults in care homes rated the food they received not only as a highlight of the day but as the most important part of the care they receive. Reflecting this importance, the Department of Health's 2002 National Minimum Standards for Care Home include those specifically relating to food and drink. The food-based standards say that care homes should:

✔ Give residents a wholesome, balanced diet in pleasing surroundings

✔ Assess new residents to assess their dietary preferences and requirements

✔ Tailor meals to stated preferences and needs

✔ Make food nutritious, attractive and appealing, using a combination of colour, texture and flavour

✔ Ensure variety for stimulation of appetite and avoid overly restricted menus – a four- to five-week cycle is ideal with popular meals provided at frequent intervals, and menus should change with the seasons

✔ Offer each resident three meals each day, at least one of which should be cooked, at intervals of less than five hours apart but at flexible times suitable to the residents

✔ Make hot and cold drinks and snacks available at all times and offer them regularly

✔ Offer a snack meal in the evening but ensure the gap between this and breakfast doesn't exceed 12 hours

✔ Make texture-modified meals attractive and appealing and well-flavoured with the items puréed and served separately

✔ Meet special dietary needs including religious and cultural requirements

✔ Inform residents of the choice of meals available to them

✔ Ensure that mealtimes are unhurried with sufficient time to finish

✔ Offer assistance discretely if necessary, and ensure that food is the correct temperature even for slow eaters and that second helpings are available without rush

As with school meals (see the earlier section 'Setting the Standard in Schools'), the Caroline Walker Trust has devised nutrient standards for daily food intakes for people in care homes that relate to fat, salt, sugar and target recommendations for nutrients such as iron or vitamin C, in which many older people are deficient. Because gaining vitamin D from diet is difficult and sun exposure may be inadequate, the guidelines also recommend a daily supplement of 10 micrograms of this vitamin. See Table 17-4.

Table 17-4 **Daily Nutritional Guidelines for Food for Older People in Residential Care Homes**

Nutrient	Level	Amount
Energy	EAR *	Women 1,810 kcal; men 2,100 kcal
Fat	35% of food energy	Women 70g; men 82g
Sugar	11% of food energy	Women 53g; men 62g
Fibre	RNI **	18g
Protein	RNI**	Women 46.5g; men 53.3g
B vitamins: B1	RNI**	Women 0.8mg; men 0.9mg

Nutrient	Level	Amount
B vitamins: B2	RNI **	Women 1.1mg; men 1.3mg
B vitamins: Niacin	RNI**	Women 12mg; men 16mg
Folate	RNI **	200mcg
Vitamin C	RNI**	40mg
Vitamin A	RNI**	Women 600 mcg; men 700mcg
Calcium	RNI	700mg
Iron	RNI**	8.7mg
Zinc	RNI**	Women 7mg; men 9.5mg
Potassium	RNI**	350mg
Salt		No more than 6g

Source: Caroline Walker Trust Expert Working Group
*Estimated Average Requirements The estimated mean or average requirement of a group for a particular nutrient. This is used for energy.
**Reference Nutrient Intakes The amount of a nutrient deemed sufficient for almost all healthy individuals in fact more than most people need. RNIs are used for protein, vitamins and minerals.

Care homes should also:

✓ **Design menus to provide the nutrients in a form liked and managed by older people.** Examples might include familiar dishes or easy to eat fruit and vegetables dishes or meat products. If salt and sugar reductions are necessary, ensure the food remains tasty given the fact that reduced taste and smell occurs in many older adults.

✓ **Consider the needs of residents who have difficulties eating.** Special cutlery or slip mats can be useful. You can also seek advice from other healthcare professionals such as occupational therapists, speech and language therapists or dietitians.

✓ **Ensure access to drinks throughout the day.** Older people have a weaker sense of thirst and residents may need encouragement to drink to reduce the risk of dehydration.

✓ **Encourage communal meals as far as possible.** Not surprisingly, studies have shown the dining environment can have a big influence of food intake in care homes. When eating with others people tend to eat more food. Attractive, inviting, well-lit and warm dining rooms with appropriate seating, tablecloths and distractions like TV and radio kept to a minimum are important.

✔ **Screen residents to monitor weight and food intake.** Homes should do this upon the resident's admission to the care home, and at intervals thereafter. Care staff should look for changes in:

• Eating or drinking less than usual, leaving more food on the plate, losing the ability to self-feed self or swallowing or chewing difficulties

• Tiredness or lack of energy or appetite, or unwanted weight loss – homes should have functioning weighing scales to monitor this

Despite the existence of standards, recent surveys from the Care Standards Commission suggest some homes aren't up to scratch and plenty of people still complain about quality and choice. To help ensure best service and to drive standards upwards, if you're considering a care home – either as a prospective resident or a relative or carer – ask to see the dining room and the menu, and enquire about the home's commitment to nutritional standards and screening as well as the budget spent on food ingredients. As a guideline you would expect this to be between at least £22-25 per person per week depending on the size of the home. Other questions you might want to ask include:

✔ Is there a varied choice of menu including foods you enjoy?

✔ Can special diets be catered for?

✔ What time are the meals served?

✔ Can you have, or make a snack or drink for yourself or your visitors?

✔ Is it possible to have meals other than at set times?

✔ Can you have an early morning cup of tea?

✔ Can you have breakfast in bed?

✔ Can you choose whether to go to the communal area or stay in your room?

Eating on the March: Nutrition in the Armed Forces

Napoleon Bonaparte famously acknowledged the importance of good nutrition to the armed forces when he said 'an army marches on its stomach'. Indeed, it was Napoleon's demands for food that was easy to carry and eat on the move for his soldiers en route to battle that led to the invention of the tin can as a way of preserving food.

Things have moved forward a lot since Napoleon's leaky tin cans, and dietitians know that the right diet is just as important for a member of the armed forces as it is for an athlete. Even with more mechanised armed forces and state-of-the-art technology, fitness and stamina are still the key to dealing with challenges met in military training and operations. Just as with sports-men and -women, the armed forces need a good diet for coping with the physical demands of training, reducing the risk of injury or illness and recovering more quickly. You need energy – in particular adequate carbohydrate – for marching and fighting, to fuel the brain to stay mentally alert and to maintain morale.

Fighting fit

The ideal diet for the armed forces remains in line with Eatwell plate – variety from all five foods – although the quantity will vary depending on activity levels.

In the past, members of the armed forces not on active duty received meals for a set fee taken in advance from their wages. In more recent times, food provision has begun to change to a 'pay as you dine' system where the onus for food choice has gone back onto the individual. The new system allows diners to just pay for what they eat. However, some are concerned that this may lead to more restricted and unhealthy food choices and a 'bag of crisps and a pasty culture', with people relying on cheap calories from snacks, takeaways and fast food eaten in their rooms rather fresh cooked meals with adequate fruit and vegetables, especially towards end of month when money may be short. So understanding how to eat healthily is more important than ever for those in the armed forces.

Here are some key tips for healthy eating for members of the armed services:

✔ Eat plenty of starchy carbohydrate foods, choosing wholegrain types where you can.

✔ Include sources of lean protein from meat, fish, eggs, beans, pulses and alternatives, and low fat dairy.

✔ Avoid fatty meat products, pies and pasties.

✔ Get your five portions of fruit and vegetables each day.

✔ Eat relatively small amounts of foods high in fat and sugar.

✔ Don't replace meals with fatty and sugary snacks or takeaways.

✔ Stay within the safe levels for alcohol (2-3 units per day for women and 3-4 units for men). For more information on alcohol units see Chapter 8.

Drinking up

Studies have suggested that members of the armed forces often fail to drink enough fluid, especially during heavy training or in the heat. Just as with athletes, fluid is vital for optimal performance – to avoid overheating, risk of injury and loss of concentration. Water on its own is fine, and so are diluted flavoured drinks, tea and coffee or sports drinks where activity is prolonged.

Supplementary information

In a recent American survey over 85 per cent of the armed personnel inter-viewed reported they used nutritional supplements – an even higher level than international athletes reported! For those eating a varied diet, extra vita-mins and minerals are both unnecessary and expensive. Similarly, you don't need protein powers and amino acids to build lean body muscle mass.

Other supplements, such as herbal tonics, caffeine and creatine, that make performance-enhancing claims offer little benefit in reality and may have unwanted side effects.

Going Beyond Porridge: Dietary Recommendations for the Prison Service

In the UK over 70,000 adults are in prison, with around another 10,000 in young offender establishments. Twenty-nine million meals are served to inmates each year with an average daily budget of approximately £2 per head.

Food in the field

Around 80 per cent of food provided to armed forces in the field is freshly cooked, but some-times those in the field have to eat from ration packs. The British army first introduced ration packs on D Day and has used them ever since, although they've undergone radical transfor-mations in recent years.

Each 24-hour operational ration pack (ORP) is designed to meet the daily needs of a man involved in active operational duty and has around 4,000 calories. Meals sealed in foil pouches are given alongside soup, savoury crackers, sweet biscuits, confectionery and puddings. More modern packs designed for hot climates include flapjacks or energy bars that don't melt, pasta, muesli, smoothies and fruit flavourings for water to encourage fluid intake.

In recent years research on young offenders in particular has started to consider whether poor diet may go beyond its influence on physical health and play a role in behaviour. Dietitians believe that poor diet can influence areas of brain development, particularly in the area known as the prefrontal cortex, which controls impulsive behaviour and is very sensitive to shortage of nutrients. Several studies, including a landmark study in Aylesbury Young Offenders' Institute in Buckinghamshire, have shown that giving inmates a multivitamin mineral and fatty acid supplement can dramatically reduce violent and antisocial behaviour. Clearly nutrition isn't the only factor influencing behaviour, but the results of studies like this have vast implications for improving the prison environment and reoffending rates.

In 2006 a National Audit Office report highlighted the fact that although over the previous ten years the quality of the diet in prisons had improved and was meeting many of the basic dietary recommendations, great scope for improvement still existed. Some of the key concerns from this report are:

✔ Inmates have high levels of energy, fat, saturated fat and salt – inappropriate for inactive adults – in the form of over reliance on pies, pastry, meat products and processed foods.

✔ Diets include insufficient fruit, vegetables, fibre and oily fish.

✔ Cooked vegetables are either held for long periods, resulting in vitamin loss, or served cold and unpalatable.

✔ There are long intervals between the evening meal and breakfast, with lack of access to snacks other than high fat, high sugar ones like confectionery.

✔ Breakfasts are inadequate – often eaten too early ifinmates are given breakfast packs the night before.

✔ Cultural and religious dietary needs aren't fully met.

✔ Vitamin D intake in meals doesn't compensate for the inmates' low exposure to sunlight.

As a result of these findings government advisory bodies have made some key recommendations for overcoming these issues. One key area of focus has been to try and improve breakfast intakes with the provision of wholegrain bread cereals, dried and fresh fruit and low fat dairy products such as milk and yoghurt. Prisons have also started to procure foods low in salt, sugar and saturated fat, improve cooking methods and restrict certain foods like chips and fatty meat produces. Many have also tried to promote healthier choices with the provision of point-of-choice information highlighting healthier choices and snacks.

In general prisons are beginning to see food as part of rehabilitation and education programmes to build not only physical but mental health of those in custody. Many are taking steps to aid adoption of healthier diets for inmates, to overcome the challenges of personal preferences and budget, and to involve everyone from governors to staff and individual prisoners.

Part IV
Processed Food

'It's the only way I can get him to
take his slimming pills.'

In this part . . .

Have you ever wondered why an originally translucent egg white turns white when you cook it? Or why frozen carrots are mushy when you defrost them? Wonder no more – just read on to find out what happens when you cook, freeze, or tin food. We also take a peek at modern food technology at its most mysterious: GM foods and additives – what do they do and how safe are they?

Chapter 18

What Is Processed Food?

S ay 'processed food' and you may think 'processed cheese'. You'd be right, of course. Processed cheese is, indeed, a processed food. But so are baked potatoes, canned tuna, frozen peas, skimmed milk, pasteurised orange juice and scrambled eggs. In broad terms, food processing is any technique that alters the natural state of food – everything from cooking to freezing to pickling to drying and more and more and more.

In this chapter you read all about how each form of processing changes food from a living thing (animal or vegetable) into an integral component of your healthy diet – and at the same time:

✔ Lengthens shelf life

✔ Reduces the risk of food-borne illnesses

✔ Maintains or improves a food's texture and flavour

✔ Upgrades the nutritional value of foods

What a set of bonuses!

Preserving Food: Five Methods of Processing

Where food is concerned, the term *natural* doesn't necessarily translate as 'safe' or 'good to eat'. Food spoils (naturally) when microbes living (naturally) on the surface of meat, a carrot, a peach or whatever reproduce (naturally) to a population level that overwhelms the food.

Sometimes (but not always) you can see, feel or smell when this is happening. You can see mould growing on cheese, feel how meat or chicken turns slippery and smell when milk turns sour. The mould on cheese, slippery slickness on the surface of the meat or chicken and odour of the milk are caused by exploding populations of microorganisms. Don't even argue with them; just throw out the food.

All food processing is designed to prevent what happens to the chicken (or the cheese or the milk). It aims to preserve food and extend its *shelf life* (the period of time when it's nutritious and safe to consume) by stemming the natural tide of biological destruction. (But wait! Not all microbes are bad news. We use 'good' ones to ferment milk to make yogurt or cheese and to produce wines and beers.)

Reducing or limiting the growth of food's natural microbe population not only lengthens its shelf life, but it also lowers the risk of food-borne illnesses. Increased food safety is a natural consequence of most processing that keeps foods usable for longer. This section discusses how food processing works.

Temperature control

Exposing food to high heat for a sufficiently long period of time reduces the natural population of bacterial spoilers and kills microbes that otherwise may make you unwell. For example, pasteurisation – heating milk or other liquids such as fruit juice to 63 degrees centigrade (145 to 154.4 degrees Fahrenheit) for 30 minutes – kills all disease-causing and most other bacteria, as does high-temperature, short-time pasteurisation (72 degrees centigrade or 161 degrees Fahrenheit for 15 seconds).

Chilling also protects food. It works by slowing the rate of microbial reproduction. For example:

- ✔ Milk refrigerated at 5 degrees centigrade (50 degrees Fahrenheit) or lower may stay fresh for almost a week because the cold prevents organisms that survived pasteurisation from reproducing.

- ✔ Fresh chicken frozen to –18 degrees centigrade (0 degrees Fahrenheit) or lower may remain safe for up to 12 months (whole) or 9 months (cut up). Most domestic freezers keep foods safe for up to three months.

Removing the water

Like all living things, the microbes on food need water to survive. Dehydrate the food and the bugs won't reproduce, which means that the food stays fresher longer. That's the rationale behind raisins, prunes and *pemmican*, a dried mix of meat, fat and berries adapted from East Coast Native Americans and served to 18th- and 19th-century sailors of every national stripe. Dehydration (loss of water) occurs when food is:

✔ Exposed to air and sunlight

✔ Heated for several hours in a very low oven or smoked (the smokehouse acts as a very low oven)

Controlling the air flow

Just as microbes need water, they also need air. Reducing the air supply almost always reduces the bacterial population. The exception is *anaerobes* (microorganisms that can live without air) such as botulinum organisms that thrive in the absence of air. But hey, every rule has an exception!

Foods are protected from air by vacuum packaging. A *vacuum* – from *vacuus*, the Latin word for empty – is a space with virtually no air. Vacuum packaging is done on a container (generally a plastic bag or a glass jar) from which the air is removed before it's sealed. When you open a vacuum-packed container, you hear a sudden little pop as the vacuum breaks.

If you don't hear a popping sound, the seal has already been broken, allowing air inside, and that means that the food inside may be spoiled or may have been tampered with. Don't taste-test: throw out the entire package, food and all.

Chemical warfare

About two dozen chemicals are used as *food additives* or *food preservatives* to prevent spoilage. (If the mere mention of chemicals or food additives makes the hair on the back of your neck rise, chill out with Chapter 20.) Here are the most common chemical preservatives:

✔ **Acidifiers.** Most microbes don't thrive in highly acid settings, so a chemical that makes a food more acidic prevents spoilage. Wine and vinegar are acidifying chemicals, and so are *citric acid*, the natural preservative in citrus fruits, and *lactic acid*, the natural acid in yogurt.

- ✔ **Bacteria busters.** Salt is *hydrophilic* (*hydro* = water, *phil* = loving). When you cover fresh meat with salt, the salt draws water up and out of the meat – and up and out of the cells of bacteria living on the meat. Ergo: the bacteria die; the meat dries. And you get to eat corned beef – which gets its name from the fact that large grains of salt were once called *corns*.

- ✔ **Mould inhibitors.** Sodium benzoate, sodium propionate and calcium propionate slow (but don't entirely stop) the growth of mould on bread. Sodium benzoate is also used to prevent the growth of moulds in cheese, margarine and syrups.

Irradiation

Irradiation is a technique that exposes food to electron beams or to *gamma radiation*, a high-energy light stronger than the X-rays your doctor uses to make a picture of your insides. *Gamma rays*, which are also known as *pico rays*, are ionising radiation, the kind that kills living cells. As a result, irradiation prolongs the shelf life of food by:

- ✔ Killing microbes and insects on plants (wheat, wheat powder, spices, dry vegetable seasonings)

- ✔ Preventing potatoes and onions from producing new sprouts at the eyes

- ✔ Slowing the rate at which some fruits ripen

- ✔ Killing disease-causing organisms such as trichinella, salmonella, E. coli and listeria (the organism responsible for a recent outbreak of food poisoning within packaged meats and cold cuts) on meat and poultry

The only irradiated foods you can buy in the United Kingdom are herbs, spices and seasoning, such as stock cubes, and these food must be labelled as such. However, in 1998 the United States Food and Drug Administration (FDA) put its stamp of approval on irradiating fresh red meat products.

For simplicity's sake, here's a list of the methods used to extend the shelf life of food. Chapters 19 and 20 explain each method in detail.

- ✔ **Air control**
  - Canning
  - Vacuum packaging
- ✔ **Chemical methods**
  - Acidification
  - Mould inhibition
  - Salting (dry salt or brine)

✔ **Irradiation**

✔ **Moisture control**

- Dehydration

- Freeze-drying, a method that combines methods of controlling the temperature, air and moisture

✔ **Temperature methods**

- Canning

- Cooking

- Freezing

- Refrigeration

Making Food Better, and Better for You

Food processing can actually make your food taste better. As if that weren't enough, it also makes a wide variety of seasonal foods (mostly fruits and vegetables) available all year long, and it enables food producers to improve the nutritional status of many basic foods by enriching or altering them to meet the needs of modern consumers.

Intensifying flavour and aroma

One advantage of commercial food processing is that it enables you to enjoy things never seen in nature – such as that ever-popular processed cheese. A more prosaic benefit of food processing is that it intensifies aroma and flavour, almost always for the better. Here's how:

✔ **Drying concentrates flavour.** A prune has a different, darker, more intensely sweet flavour than a fresh plum. On the other hand, dried food can be hard and tough to chew and you may need to rehydrate the food before eating it. (Have you ever tried a dried mushroom?)

✔ **Heating heightens aroma by quickening the movement of aroma molecules.** In fact, your first tantalising hint of dinner is usually the scent of cooking food. Chilling has the opposite effect: it slows the movement of the molecules. To sense the difference, sniff a plate of cold roast beef versus hot roast beef straight from the oven. Or sniff two glasses of vodka, one warm, one icy from the freezer. One comes up scent free; the other has the olfactory allure of pure fresh coffee. Guess which is which. Or you can pass up the guessing and try for yourself. Nothing like first-hand experience!

✔ **Warming foods also intensifies flavours.** This development is sometimes beneficial (warm roast beef is somehow more savoury than cold roast beef), sometimes not (warm milk is definitely not as popular as the icy-cold version).

✔ **Changing the temperature also changes texture.** Heating softens some foods (potatoes are a good example) and solidifies others (think eggs). Chilling keeps the fats in pâté firm so the stuff doesn't melt down into a puddle on the plate. Ditto for the gelatine that keeps jelly and blancmange standing upright.

Adding nutrients

The addition of vitamins and minerals to basic foods has helped eliminate many once-widespread nutritional deficiency diseases. The practice is so common that you take the following for granted:

✔ White flour is enriched with calcium and some breakfast cereals are given extra B vitamins to replace the vitamins lost when whole grains are stripped of their nutrient-rich covering to make white flour or cereal.

✔ Some breakfast cereals are also given iron to replace what's lost in milling and to make it easier for you to reach the RNI (reference nutrient intake) for this important mineral.

✔ Margarine has added vitamin D to reduce the risk of the bone-deforming vitamin D-deficiency diseases rickets (among children) and osteomalacia (among adults).

Combining benefits

Adding genes from one food (such as corn) to another food (such as tomatoes) may make the second food taste better and stay fresh longer. Check out Chapter 20 for more about genetic engineering at the dinner table.

Making foods functional

Functional foods are big business and more and more foods with supposed 'special' properties hit the shelves every day. The official definition of functional foods is foods that provide benefits beyond their basic nutritional properties that therefore may provide medical and health benefits.

Although you can consider many foods in their natural state to be functional foods – for example, fresh fruit and vegetables, breast milk, whole grains and soya products – the concept of functional foods is rapidly being extended to commercially manufactured products. Cereals, yoghurt drinks and other dairy foods, spreads and even confectionary have functional ingredients added to make them that bit better for you and make you more likely to buy the products.

Of course, some very good functional foods exist: eggs with additional omega-3 fatty acids, yoghurt drinks with probiotics and cereals with prebiotic fibre added to them. But in our view, rules should govern which foods can be made functional and which can't, but at the moment it's down to food manufacturers to decide. This leaves you having to use your common sense when you try to decide whether a probiotic choclate bar is really going to any better for you than an ordinary bar. Just remember, if something sounds too good to be true, it probably is!

Faking It: Alternative Foods

In addition to its many other benefits, food processing offers you some totally fake but widely appreciated substitute fats and sweeteners. Actually, these may be just the tip of the iceberg, so to speak. Quorn, food made from fungi – yes, fungi – is widely available now as a low-fat/high-fibre meat substitute. For example, Quorn's popularity is likely to increase as more and more people become interested in reducing their red meat and saturated fat intakes.

Fake fats

Fat carries desirable flavours and gives food a nice feeling in the mouth. However, fat is also high in calories, and some fats (the saturated kind we describe in Chapter 5) can clog your arteries. One way to deal with this problem is to eliminate the fat from food (see milk, in the previous section) or to follow a very low-fat diet, which can be difficult to stick to long term. Another solution is to head for the food lab and create a no- or low-calorie, non-clogging substitute or fake fat.

Olestra/Olean

Olestra/Olean is a no-calorie chemical from sugar and vegetable oils used in the past in the United Kingdom in snack foods such as potato crisps. Olestra is indigestible, which means that it adds no calories, fat or cholesterol to food. Unfortunately, as Olestra speeds through your intestinal tract, it's likely to pick up and swoosh along some fat-soluble nutrients such as vitamin A, vitamin D, vitamin E and cancer-fighting carotenoids, as well as producing unpleasant side effects for some people including anal leakage – yuk! Although Olestra has been approved for use in the United States since 1996, UK manufacturers don't use the chemical in products. For the moment the best way to avoid too much fat in your diet is – yes, you've guessed it – eat a balanced diet with only moderate amounts of fats and fatty foods.

Simplesse

Simplesse is a low-calorie fat substitute made by heating and blending proteins from egg whites and/or milk into extremely tiny round balls that taste like fat. Simplesse has 1 to 2 calories per gram versus 9 calories per gram for real fats or oils. Simplesse isn't used in the United Kingdom and even in the United States it's not recommended for young children because they need essential fatty acids found in real fats. Simplesse may be problematic for people who are:

✔ Sensitive to milk

✔ Sensitive to eggs

✔ On low-protein diets (for example, kidney disease patients)

Substitute sweeteners

Various substances are used as alternative sweeteners to sucrose (table sugar), which is full of calories. They fall into two categories:

✔ **Intense (non-nutritive) sweeteners:** Compounds that have the unique property of giving an intense sweetness in minute quantities. The energy or calorie contribution to food is negligible because intense sweeteners are used in such small amounts.

✔ **Nutritive (bulk) sweeteners:** Sugar substitutes used in similar quantities as sucrose; they add bulk and have some of the properties of sucrose. All add some energy and carbohydrate value to food, but some provide less than sucrose.

Some sweeteners are used as tabletop sweeteners, to add to your tea for example, and others are only found in manufactured products such as sweets and cakes. Intense sweeteners are controlled by The Sweeteners in Foods Regulations 1995 (amended 1997), which follow European Commission directives. These regulations specify the types of foods in which sweeteners can be used, set maximum levels at which sweeteners can be used and monitor levels of intake to ensure that they don't exceed the acceptable daily intake (ADI).

Manufacturers now use intense sweeteners in so many foods from drinks to chewing gum that it's easy for some people, especially children and people with diabetes, to exceed the ADI. If you consume a lot of foods sweetened with intense sweeteners, vary the varieties of products you eat to avoid exceeding the ADI of a particular intense sweetener.

The best-known substitute sweeteners (in order of their discovery) are:

- **Saccharin (E954):** This synthetic sweetener was discovered at Johns Hopkins University in the United States in 1879. A ban on saccharin was proposed in 1977, after it was linked to bladder cancer in rats; however, it's still on the market, and diabetics who've used saccharin for years show no excess levels of bladder cancer. A health warning label appears with saccharin-sweetened products in the United States indicating that it's a mild rodent carcinogen. The maximum ADI for saccharin is 5 milligrams per kilogram of body weight (about 20 saccharin tablets a day for a 60 kilogram adult; less for children). This level is pretty easy to exceed if you have a couple of diet drinks a day, and add saccharin to your tea and coffee.

- **Cyclamates (E952):** These sweeteners surfaced in 1937 at the University of Illinois. They were tied to cancer in laboratory animals and banned in the UK until 1996. No ill effects in human beings have been attributed to cyclamates, which are available for use as a tabletop sweetener. However, as with saccharin, if you consume lots of soft drinks, be careful not to exceed the ADI of only 1.5 milligrams per kilogram of body weight.

- **Aspartame (E951):** Available in the United Kingdom since 1983, *aspartame* is a combination of two amino acids, aspartic acid and phenylalanine. The problem with aspartame is that although it doesn't have the bitter aftertaste of saccharin, it's much less stable and during digestion it breaks down into its constituent ingredients. The same thing happens when aspartame is exposed to heat. That's trouble for people born with *phenylketonuria* (PKU), a metabolic defect characterised by a lack of the enzyme needed to digest phenylalanine. The excess amino acid can pile up in brain and nerve tissue, leading to mental retardation in young children. In the United Kingdom foods and drinks containing aspartame carry a warning to this effect. A similar additive called salt of aspartame (acesulfame, E962) is currently used in the UK under a temporary two-year authorisation. We wait to see whether it becomes a permanent fixture in our foods.

- **Sucralose (E955):** Discovered in 1976, sucralose is the only no-calorie sweetener made from sugar. Your body doesn't recognise it as a carbohydrate or a sugar, so it zips through your intestinal tract unchanged. More than 100 scientific studies conducted during a 20-year period attest to its safety, and the United States FDA recently approved its use in a variety of foods, including baked goods, sweets, substitute dairy products and frozen desserts. Sucralose is now available in the UK under the brand name of Splenda.

✔ **Acesulfame-K (E950):** The *K* is the chemical symbol for potassium, and you find this artificial sweetener, with a chemical structure similar to saccharin, in baked goods, chewing gum and other food products. Acesulfame-K became available in the United Kingdom in 1983.

✔ **Thaumatin (E957):** This protein is extracted from the seeds of a West African plant. Thaumatin is far too concentrated in sweetness to be used as a tabletop sweetener and is used as a flavour enhancer or combined with other sweeteners.

✔ **Neohesperidine DC (NHDC) (E959):** The most recently approved intense sweetener in the United Kingdom is used more and more in soft drinks, baked goods, confectionery, savoury snacks and alcohol-free beer. Made from two protein compounds, this intense sweetener hasn't yet been approved for use throughout the European Union.

Table 18-1 compares the calorie content and sweetening power of sugar versus substitute sweeteners. For comparison, sugar has 4 calories a gram.

Table 18-1	Comparing Substitute Sweeteners	
Sweetener	*Calories per Gram*	*Sweetness Relative to Sugar* (Sucrose)*
Acesulfame-K	0	150 times sweeter than sugar
Aspartame	4**	200 times sweeter than sugar
Cyclamates	0	30–60 times sweeter than sugar
Neohesperidine	0	2000 times sweeter than sugar
Sugar (sucrose)	4	
Saccharin	0	300 times sweeter than sugar
Sucralose	0	600 times sweeter than sugar
Thaumatin	0	2000–2500 times sweeter than sugar

*The range of sweetness reflects estimates from several sources.
**Aspartame has 4 calories per gram, but you need so little to get a sweet flavour that you can count it as 0 calories per serving.

Follow that bird

You can sum up the essence of food processing by following the trail of one chicken from the farm to your table. (Vegetarians are excused from this section.)

A chicken's first brush with processing comes right after slaughtering. It's plucked, cut into pieces and packed off to the food factory. It travels on ice to slow the natural bacterial decomposition. In the food factory, your chicken may be boiled and tinned in large pieces, or boiled and cut up and tinned in small portions like tuna fish, or boiled into chicken soup to be tinned, or dehydrated into stock cubes, or cooked with veggies and tinned as chicken stew, or fried and frozen in whole pieces, or roasted, sliced and frozen into a chicken dinner, or . . . you get the picture.

When you buy a fresh (raw) chicken instead of a cooked one, you perform similar rituals in your own kitchen. First, the chicken goes to the refrigerator (or freezer), then to the oven for thorough cooking to make sure that no stray bacteria contaminates your dinner table (or you) and then back to the fridge for the leftovers. In the end, the chicken's been processed. And you've eaten. That's food processing.

Chapter 19

Cooking and Keeping Food

*Y*ou can bet that the first cooked dinner was an accident involving some poor wandering animal and a flash of lightning that instantly charred the beast into medium sirloin. Then a caveman, attracted by the aroma, tore off a sizzled chunk, and mealtimes were never the same again.

After that, it was but a hop, a skip and a jump, anthropologically speaking, to gas cookers, electric grills and microwave ovens. This chapter explains how these handy technologies affect the safety, nutritional value, appearance, flavour and aroma of the foods that you heat and eat.

What's Cooking?

Ever since humans discovered fire and how to control cooking – rather than having to wait for a passing flash of lightning – the human race has generally relied on three simple ways of heating food:

- ✔ **An open flame:** You hold the food directly over – or under – the flame or put the food on a hob on top of the flame. The electric heating coil is a 20th-century variation on the open flame.

- ✔ **A closed box:** You put the food in a closed box (an oven) and heat the air in the oven to create high-temperature dry heat.

- ✔ **Hot liquid:** You submerge the food in hot liquid or suspend the food over the liquid so that it cooks in the steam escaping from the surface.

Cooking food in a wrapper such as aluminum foil combines two methods – an open flame (the grill) or a closed box (the oven) plus the steam from the food's own juices (hot liquid).

Here are the basic methods used to cook food with heat generated by fire or an electric coil:

Open Flame	*Hot Air*	*Hot Liquid*
Frying	Baking	Boiling
Grilling	Roasting	Deep-frying
Toasting		Poaching
		Simmering
		Steaming
		Stewing

Converting between Fahrenheit and centigrade

Pssst! Here's how to convert temperatures from Fahrenheit (F) to Centigrade (C) and back again:

1. Degrees centigrade =
$$\frac{(\text{degrees F} - 32)}{9} \times 5$$

For example, to convert the Fahrenheit boiling point of water (212 degrees F) to the centigrade boiling point of water (100 degrees C):

$$\frac{(212 - 32) \times 5}{9} = 100$$

2. Degrees Fahrenheit =
$$\frac{(\text{degrees C}) \times 9}{5} + 32$$

For example, to convert the centigrade boiling point of water (100 degrees C) to the Fahrenheit boiling point of water (212 degrees F):

$$\frac{100 \times 9}{5} + 32 = 212$$

Cooking with electromagnetic waves

A gas or electric oven generates thermal energy (heat) that warms and cooks food. A microwave oven generates electromagnetic energy (microwaves) produced by a device called a magnetron (see Figure 19-1).

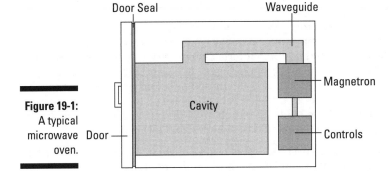

Door Seal   Waveguide

Magnetron

Cavity

Controls

Figure 19-1: A typical microwave oven.   Door

Microwaves transmit energy that excites water molecules in food. The water molecules leap about like hyperactive 3-year-olds, producing heat that cooks the food. The dish holding food in a microwave oven generally stays cool because it has so few water molecules.

Some people are concerned that the microwave is an unhealthy method of cooking. In fact, cooking vegetables in the microwave is one of the best methods of retaining vitamins and minerals – the microwave is a great kitchen asset.

Never use a metal container to heat food in a microwave.

Cooking away contaminants

Many microorganisms that live naturally in food are harmless or even beneficial. For example:

- ✔ *Lactobacilli* (*lacto* = milk; *bacilli* = rod-shaped bacteria) digest sugars in milk and convert the milk to yogurt.
- ✔ Nontoxic moulds convert milk to blue cheese: The blue ribbons in the cheese are mould.

Some organisms, however, carry the risk of food poisoning. Every year 5.5 million people in the United Kingdom experience diarrhoea – a common symptom of food poisoning – after eating food contaminated with such an organism, but only about 15 per cent of these people ever report their symptoms to their doctor, presumably thinking they've 'just picked up a bug'. Well they have – but from their own kitchen or someone else's!

In extreme cases food poisoning can lead to serious illness and even death.

- ✔ During pregnancy, a foetus whose mother consumes listeria-contaminated food may suffer damage or, in extreme cases, may die. Listeria is caused by the bacteria *Listeria monocytogenes*. Pregnant women should avoid soft-ripened cheese like Camembert and mould-ripened cheese like Stilton. Unpasteurised milk and pre-prepared or pre-cooked foods that don't need reheating are all foods that can contain high levels of the bacteria, which most people don't suffer ill effects from. They also should avoid pate which can contain listeria.

- ✔ In 1997 the world's worst food poisoning outbreak happened in Wishaw, Scotland, when 160 people were admitted to hospital and 20 died after eating meat that had been cooked and packaged in machinery contaminated with the deadly bacteria *Escherichia coli 0157: H7* (E. coli 0157).

- ✔ The toxin produced by *Clostridium botulinum*, which thrives in the absence of air (and therefore in low-acid, tinned food), is potentially fatal. Your only warning of its presence may be a swollen tin – which should be discarded without ever being opened.

- ✔ Raw meat, poultry and unpasteurised milk contaminated with *Campylobacterium jejuni* (C. jejuni) have been linked to Guillain-Barré syndrome, a paralytic illness that sometimes follows flu infection.

These bacteria are most dangerous for the very young, the very old and those whose immune systems have been weakened by illness or medication.

Cooking food thoroughly and keeping it hot (or chilling it quickly) after you cook it destroys many dangerous bugs or slows the rate at which they reproduce, thus reducing the risk of food poisoning. Table 19-1 lists some common *pathogens* (disease-causing organisms) linked to food-borne illnesses and notes the foods most likely to harbour them.

Table 19-1	Disease-Causing Organisms in Food
The Bug	*Foods Where You Are Most Likely To Find It*
Campylobacterium jejuni	Raw meat and poultry, unpasteurised milk
Clostridium botulinum	Tinned low-acid foods, vacuum-packed smoked fish and meat

The Bug	*Foods Where You Are Most Likely To Find It*
Clostridium perfringens	Foods made from poultry or meat
Escherichia coli	Raw beef
Listeria monocytogenes	Raw meat and seafood, raw milk, some raw cheeses
Salmonella bacteria	Poultry, meat, eggs, dairy products
Staphyloccus aureaus	Custards and cream and milk desserts; egg, chicken, and tuna salads; prepared meat dishes
Bacillus cereus	Rice

Table 19-2 shows the recommended safe cooking temperatures for various foods. The table shows the *internal temperature*. Some foods can seem very hot on the outside but may not be hot enough to kill bacteria on the inside. You can buy thermometers to measure the internal temperature of foods from hardware and kitchen shops.

Table 19-2	**How Hot Is Safe?**
This Food . . .	*Is Generally Safe to Eat When Cooked to This Internal Temperature*
Eggs and egg dishes	
Eggs	Cook until yolk and white are firm
Egg dishes	160°F (71°C)
Minced meat and meat mixtures	
Turkey, chicken	165°F (74°C)
Veal, beef, lamb, pork*	165°F (74°C)
Fresh beef*	
Medium rare	145°F (63°C)
Medium	160°F (71°C)
Well done	170°F (77°C)
Fresh pork	
Medium	160°F (71°C)
Well done	170°F (77°C)

(continued)

Table 19-2 *(continued)*

This Food . . .	Is Generally Safe to Eat When Cooked to This Internal Temperature
Poultry	
Chicken, whole	180°F (82°C)
Turkey, whole	180°F (82°C)
Poultry breasts, roasts	170°F (77°C)
Poultry thighs, wings	Cook until juices run clear
Stuffing (cooked alone or in bird)	165°F (74°C)
Duck and goose	180°F (82°C)
Fresh (raw)	160°F (71°C)
Precooked (to reheat)	140°F (60°C)

Undercooked hamburger is a major source of the potentially lethal organism E. coli. To be safe, the internal temperature must read 165°F.
Adapted from USDA Food Safety and Inspection Service, 'A Quick Consumer Guide to Safe Food Handling', Home and Garden Bulletin, No. 248 (August 1995).

Exploring How Cooking Affects Food

Cooking food changes the way it feels, looks, tastes and smells. This section digs deeper into how cooking affects different nutrients.

Two hours – and you're out!

Microorganisms thrive on food at temperatures between 40 and 140 degrees Fahrenheit (4.4 and 60 degrees centigrade). The cooking temperature that inactivates many, though not all, bad guys is 140 degrees Fahrenheit. For maximum safety, follow the two-hour rule: never allow food to sit at temperatures between 40 and 140 degrees Fahrenheit for more than two hours.

More questions about food safety?

✔ Check out the Food Standards Agency advice at www.eatwell.gov.uk.

✔ Contact your local Environmental Health Department (look in your Yellow Pages).

Cook me tender: Changing texture

Exposure to heat alters the structures of proteins, fats and carbohydrates, so it changes food's *texture* (the way food particles are linked to make the food feel hard or soft). In other words, cooking can turn crisp carrots mushy and soft steak to shoe leather.

Protein

Proteins are made of very long molecules that sometimes fold over into accordion-like structures (see Chapter 4 for details about proteins). Although heating food doesn't lower its protein value, it does:

✔ Break protein molecules into smaller fragments

✔ Cause protein molecules to unfold and form new bonds to other protein molecules

✔ Make proteins clump together

This process is called *coagulation*. Need an example? Think of an egg. When you cook one, the long protein molecules in the white unfold, form new connections to other protein molecules and link up in a network that tightens to squeeze out moisture so that the egg white hardens and turns opaque. The same unfold–link–squeeze reaction turns translucent poultry firm and white and makes gelatine set. The longer you heat proteins, the stronger the network becomes, and the tougher, or more solid, the food will be.

To see this work, scramble two eggs – one beaten and cooked plain, and one beaten with milk and then cooked. Adding liquid (milk) makes squeezing out all the moisture more difficult for the protein network. So the egg with the added milk cooks up softer than the plain egg.

Fat

Heat melts fat, which can run off food, lowering the calorie count. In addition, cooking breaks down connective tissue – the supporting framework of the body, which includes some adipose (fatty) tissue – thus making the food softer and more pliable. You can see this most clearly when cooking fish. The fish flakes when it's done because its connective tissue has been destroyed.

When meat and poultry are stored after cooking, their fats continue to change, this time by picking up oxygen from the air. Oxidised fats have a slightly rancid taste. You can slow – but not entirely prevent – this reaction by cooking and storing meat, fish and poultry under a blanket of food rich in *antioxidants*, chemicals that prevent other chemicals from reacting with oxygen. Vitamin C is a natural antioxidant, so gravies and marinades made with tomatoes, citrus fruits or tart cherries slow the natural oxidation of fats.

Grains: Split personality performers

In cooking, grains such as corn exhibit split personalities – part protein, part complex carbohydrates. When you boil a corn on the cob, the protein molecules inside the kernels do the unfold–link–squeeze dance. At the same time, carbohydrate starch granules begin absorbing moisture and then soften.

The trick to boiling perfect corn is controlling this process, removing the corn from the water when starch granules have absorbed enough moisture to soften the kernels but before the protein network has tightened. That's why cookbooks advise a short stay in the saucepan.

Carbohydrates

Cooking has different effects on simple carbohydrates and complex ones (if you're confused about carbohydrates, see Chapter 7). When heated:

✔ Simple sugars – such as sucrose or the sugars on the surface of meat and poultry – caramelise, or melt and turn brown. (Think of crème brûlée.)

✔ Starch, a complex carbohydrate, becomes more absorbent, which is why pasta expands and softens in boiling water.

✔ Some dietary fibres (gums, pectins, hemicellulose) dissolve, so vegetables and fruits soften when cooked.

The last two reactions – absorption and dissolved cell walls – can improve the nutritional value of foods by making the nutrients inside previously fibre-stiffened cells more available to your body.

Another, less beneficial effect that heat has on carbs was reported in 2002 in a study conducted at Stockholm University. The study shows that exposing high-carb foods, such as potatoes and grains, to very high cooking temps (for instance making potatoes into crisps, or flour into bread) triggers the production of odourless white crystals of *acrylamide*, a chemical known to cause cancer in animals and possibly linked with cancer in humans. At a follow-up United Nations-sponsored meeting in Geneva, international food scientists called for further research. The Food Standards Agency is taking this research seriously and is monitoring the results of studies currently being carried out throughout the world. Right now, no evidence suggests that the amounts of acrylamide in crisps and bread pose a serious threat to human health.

Enhancing flavour and aroma

Heat degrades (breaks apart) flavour and aroma chemicals. As a result, most cooked food has a more intense flavour and aroma than raw food.

A good example comes from the mustard oils that give cruciferous vegetables, such as cabbage and cauliflower, their distinctive (some may say offensive) odours. The longer you cook these vegetables, the worse they smell. On the other hand, heat destroys *diallyl disulfide*, which is the chemical that gives raw garlic its unmistakable taste and smell. So cooked garlic tastes and smells milder than the raw version.

Altering the palette: Food colour

Carotenoids – the natural red and yellow pigments that make carrots orange, sweet potatoes orange and tomatoes red – are practically impervious to heat and the acidity or alkalinity of cooking liquids. No matter how you cook them or how long, carotenoids stay bright and sunny.

You can't say the same for the other pigments in food: the ones that make food naturally red, green or white react – usually for the worse – to heat, acids (such as wine, vinegar or tomato juice) and basic (alkaline) chemicals (such as mineral water, or baking soda and water). Here's a brief rundown on the colour changes that you can expect when you cook food.

- Beetroot and cabbage get their colour from pigments called *anythocyanins*. Acids make these pigments redder. Alkaline solutions fade anythocyanins from red to bluish purple.

- Potatoes, cauliflower, rice and white onions are white because they contain white pigments called *anthoxanthins*. When anthoxanthins are exposed to alkaline chemicals (mineralised water or baking soda), they turn yellow or brownish. Acids prevent this reaction.

- Green veggies are coloured by *chlorophyll*, a pigment that reacts with acids in cooking water (or in the vegetable itself) to form *pheophytin*, a brown pigment. Fast cooking at high heat lessens these colour changes.

- The natural red colour of fresh meat comes from *myoglobin* and *haemoglobin* in blood. When meat is heated, the pigment molecules are *denatured*, or broken into fragments. They lose oxygen and turn brown or – after long cooking – develop the really unappetising grey characteristic of over-cooked boiled meats. This inevitable change is more noticeable in beef than in pork or veal because beef starts out naturally redder.

A hot story

When you heat fats, their molecules break apart into chemicals known as *free radicals*, molecule fragments that may hook up together to form potentially carcinogenic (cancer-causing) compounds. These compounds are produced in higher numbers at higher heats; the usual safe cut off is around 500 degrees Fahrenheit (260 degrees centigrade). Burned fat or smoking oil, for example, has more nasties than plain melted fat or oil that's warm but not smoking.

As a result, many nutritionists warn against eating the crisp, crinkly, absolutely yummy browned top layer of foods, especially burned and barbecued meats, which in 1998 were tentatively linked to a higher risk of breast cancer in women. Of course, the theory has yet to be proven, and as is true with so much in modern nutrition, the story may be more complicated than it seems at first glance.

Choosing Cookware: How Pots and Pans Affect Food

A pot is a pot is a pot, right? No way! In fact, your choice of pots can affect the nutrient value of food by:

- Adding nutrients to the food
- Slowing the natural loss of nutrients during cooking
- Actively increasing the loss of nutrients during cooking

In addition, some pots make the food's natural flavours and aromas more intense, which, in turn, can make the food more – or less – appetising. Read on to find out how your pot can change your food. And vice versa.

Aluminium

Aluminium is lightweight and conducts heat well. That's good. But the metal:

- Makes some aroma chemicals smellier (particularly those in the cruciferous vegetables like cabbage, broccoli and Brussels sprouts)
- Flakes off, turning white foods (such as cauliflower or potatoes) yellow or grey

Aluminium flaking isn't hazardous to your health: Cooking with aluminium pans doesn't increase your risk of developing Alzheimer's disease, as some rumours have it. True, cooking salty or acidic foods (wine, tomatoes) in aluminium pans increases the flaking, but even then, the amount of aluminium you get from the pan is less than you get naturally every day from food and water (see Chapter 10 for the lowdown on aluminium as a natural nutrient).

Copper

Copper pots heat steadily and evenly. To take advantage of this property, many aluminium or stainless-steel pots are made with a layer of copper sandwiched into the bottom. But naked copper is a potentially poisonous metal. That's why copper pots are lined with tin or stainless steel. Whenever you cook with copper, periodically check the lining of the pan. If it's damaged – meaning that you can see the orange copper peeking through the silvery lining – have the pan relined or throw it out.

Ceramics

The chief virtue of plain terra cotta (the orange clay that looks like red bricks) is its *porosity*, a fancy way of saying that terra cotta roasting and baking pans allow excess steam to escape while holding in just enough moisture to make bread so moist and chicken such tender pickings.

Decorated ceramic vessels are another matter. For one thing, the glaze makes the pot much less porous, so that meat or poultry cooked in a covered painted ceramic pan steams instead of roasts. The practical result – a soggy surface rather than a crisp one. More important, some pigments used to paint or glaze the pots contain lead. To seal the decoration and prevent lead from leaching into food, the painted pots are *fired* (baked in an oven). If the pots are fired in an oven that isn't hot enough or if they're not fired for a long enough period of time, lead will leach from ceramics when in contact with acidic foods such as fruit juices or foods marinated in wine or vinegar.

Copper bottoms and egg whites: A chemical team

When you whip an egg white, its proteins unfold, form new bonds and create a network that holds air in. That's why the runny white turns into stable foam.

You can certainly whip egg whites successfully in a glass or ceramic bowl – chilled, and absolutely free of any fat, including egg yolk, which would prevent the proteins from linking tightly. But the best choice is copper: the ions (particles) flaking off the surface bind with and stabilise the foam. (Aluminium ions stabilise but darken the whites.)

But wait. Didn't we just say copper is toxic? Yes, but the amount you get in an occasional batch of whites is so small it's insignificant, safety-wise.

Ceramics made in the United Kingdom, the United States and Japan are generally considered safe, but for maximum protection, hedge your bets. Unless the pot comes with a tag or brochure that specifically says it's acid safe, don't use it for cooking or storing foods. And always wash decorated ceramics by hand; repeated passes through the dishwasher can wear down the surface.

Enamelware

Enamelled pots are made of metal covered with *porcelain*, a fine translucent china. Enamelware heats more slowly and less evenly than plain metal. A good-quality enameled surface resists discolouration and doesn't react with food. But it can chip, and it's easily marked or scratched by cooking utensils other than wood or hard plastic. If the surface chips and you can see the metal underneath, discard the pot to prevent metal flakes escaping into your food.

Glass

Glass is a neutral material that doesn't react with food.

Don't use glass-and-metal pots in the microwave oven. The metal blocks microwaves. More important, it can cause *arcing* – a sudden electrical flare that may damage the oven and scare you out of your wits.

Iron cookware

Like aluminium, iron pots are a good-news/bad-news item. Iron conducts heat well and stays hot significantly longer than other pots. It's easy to clean. It lasts forever, and it releases iron ions into food, which may improve the nutritional value of dinner. Nutritionists discovered that beef stew (0.7 milligrams of iron per 100 grams, raw) can end up with as much as 3.4 milligrams of iron per 100 grams after cooking slightly longer than an hour in an iron pot.

Don't rush off to buy a set of iron pans yet. The iron that flakes off the pot may be a form of the mineral that your body can't absorb. Also, more iron isn't necessarily better. It encourages oxidation (bad for your body) and can contribute to excess iron storage in people who have *haemochromatosis*, a condition that leads to iron build-up that may damage internal organs.

By the way, did we mention that pumping iron isn't a bad way to describe the experience of cooking with iron pots? They're really, really heavy.

Nonstick

Nonstick surfaces are made of plastic (*polytetrafluoroethylene* to be exact; PTFE for short) plus hardeners – chemicals that make the surface, well, hard. As long as the surface is unscratched and intact, the nonstick surface doesn't react with food.

Nonstick pans are a dieter's delight. They enable you to cook without added fat, but using them may also lighten your wallet. They scratch easily. Unless you stick scrupulously to wooden or plastic spoons, your pan can end up looking like chickens have been pecking at the surface. The good news is that scratched nonstick pots and pans aren't a health hazard. If you swallow tiny pieces of the nonstick coating, they pass through your body undigested.

However, when nonstick surfaces get very hot, they may:

✔ Separate from the metal to which they're bound (the sides and bottom of the pan).

✔ Emit odourless fumes. If the cooking area isn't properly ventilated, you may experience *polymer fume fever* – flu-like symptoms with no known long-term effect. To prevent this, keep the hob flame moderate and the windows open.

Stainless steel

Stainless steel is an *alloy* (a substance composed of two or more metals), mostly made of iron. The virtues of stainless steel are hardness and durability. The drawback is poor heat conduction. In addition, the alloy includes nickel, a metal to which many people are sensitive. If your stainless steel pot is scratched deeply enough to expose the inner layer under the shiny surface, discard it.

Stainless doesn't always mean stainless

The absolute truth is that stainless steel isn't really stainless, after all. Check this out:

✔ When exposed to high heat, stainless steel develops a characteristic multihued 'rainbow' discoloration.

✔ Starchy foods (pasta, potatoes) may darken the pot.

✔ Undissolved salt can pit the surface.

Sorry to burst your bubble, but stainless isn't blameless!

Plastic and paper

Plastic melts and paper burns, so you obviously can't use plastic or paper containers in a conventional oven. But can you use them in the microwave? No problem! As long as you pick a proper plastic.

When plastic dishes or plastic wrap are heated in a microwave oven, they may emit potentially carcinogenic compounds that can migrate into your food. To reduce your exposure to these compounds, the Food Standards Agency (FSA) says you need to choose plastic containers labelled 'Suitable for microwave oven use'. Thin plastic storage bags and margarine tubs are convenient but way, way off-limits. Here are three common-sense tips for using the right kind of plastics in the microwave:

- ✔ Follow the directions on the plastic container or package. If it doesn't say 'microwaveable', it isn't. For example, polystyrene and other take-away food containers are rarely microwaveable, so put the food into a different container before reheating it.

- ✔ Trays for microwave meals are meant to be used only once; after you heat the food, chuck the tray.

- ✔ When covering food to prevent splatters, use microwave-safe plastic wraps or clingfilm only.

Microwave-safe plastics are required to meet strict safety standards, and repeated studies show no ill effects from their minimal leakage. On the other hand, if even very small exposure makes you edgy, you can switch to glass or ceramic dishes specifically made for use in microwave ovens. Splatter-proof the dish with waxed paper, parchment paper or white paper towels labelled safe for microwave use.

Protecting the Nutrients in Cooked Foods

Myth: All raw foods are more nutritious than cooked ones.

Fact: Some foods (such as meat, poultry and eggs) can be positively dangerous when consumed while raw (or undercooked). Other foods are less nutritious raw because they contain substances that destroy or disarm other nutrients. For example:

- ✔ Red cabbage, Brussels sprouts, blueberries and blackberries contain an enzyme that destroys thiamin (vitamin B1). Heating the food inactivates the enzyme.

- ✔ Raw beans, legumes and peanuts contain enzyme inhibitors that interfere with the work of enzymes that enable your body to digest protein. Heating disarms the enzyme inhibitor.

But there's no denying that some nutrients are lost when foods are cooked. Simple strategies such as steaming food rather than boiling or microwaving significantly reduce the loss of nutrients when you're cooking food.

Maintaining minerals

Virtually all minerals are unaffected by heat. Cooked or raw, food has the same amount of calcium, phosphorus, magnesium, iron, zinc, iodine, selenium, copper, manganese, chromium and sodium. The single exception to this rule is potassium, which – although not affected by heat or air – escapes from foods into the cooking liquid.

Those volatile vitamins

With the exception of vitamin K and the B vitamin niacin, which are very stable in food, many vitamins are sensitive flowers that are easily destroyed when exposed to heat, air, water or fats (cooking oils). Table 19-3 shows which nutrients are sensitive to these influences.

Table 19-3		What Takes Nutrients Out of Food?		
Nutrient	*Heat*	*Air*	*Water*	*Fat*
Biotin			X	
Folate	X	X		
Pantothenic acid	X			
Potassium			X	
Riboflavin			X	
Thiamin	X		X	
Vitamin A	X			X
Vitamin B6	X	X	X	
Vitamin B12	X		X	
Vitamin C	X	X	X	
Vitamin D				X
Vitamin E	X	X		X
Vitamin B6	X	X	X	

To avoid specific types of vitamin loss, keep in mind the following tips:

✔ **B vitamins:** Strategies that conserve protein in meat and poultry during cooking also work to conserve the B vitamins that leak out into cooking liquid or drippings: Use the cooking liquid in soup or sauce. ***Caution:*** Don't shorten cooking times or use lower temperatures to lessen the loss of heat-sensitive vitamin B12 from meat, fish or poultry. These foods and their drippings or juices must be thoroughly cooked to ensure that they're safe to eat.

Don't rinse grains (rice) before cooking, unless the package advises you to do so (some rice does need to be rinsed). Washing rice once may take away as much as 25 per cent of the thiamin (vitamin B1). Bake or toast cakes and breads only until the crust is light brown to preserve heat-sensitive Bs.

✔ **Vitamins A, E and D:** To reduce the loss of fat-soluble vitamins A and E, cook with very little oil. For example, bake or grill vitamin A-rich liver oil free instead of frying. Ditto for vitamin D-rich fish.

✔ **Vitamin C:** To reduce the loss of water-soluble, oxygen-sensitive vitamin C, cook fruits and vegetables in the least possible amount of water. For example, when you cook cabbage covered in water, the leaves lose as much as 90 per cent of their vitamin C. Reverse this and cook cabbage in about 2.5 centimetres (1 inch) of water and you hold on to more than 50 per cent of the vitamin C.

✔ Serve cooked vegetables quickly: after 24 hours in the fridge, vegetables lose a quarter of their vitamin C; after two days, nearly half.

✔ Root vegetables (carrots, potatoes, sweet potatoes) baked or boiled whole, in their skins, retain practically all of their vitamin C.

As a general rule, when cooking vegetables grown above the ground, add the vegetables to boiling water. For vegetables grown under the ground, add the vegetables to cold water and bring to the boil.

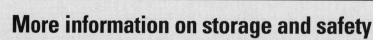

More information on storage and safety

Want to find out more about food storage and safety? Contact one of these agencies:

✔ Food Standards Agency (FSA, the Government food safety watchdog): `www.food.gov.uk`

✔ Foodlink (food safety advice from the Food and Drink Federation): `www.foodlink.org.uk`

Keeping Food Fresh

Cold air, hot air, no air and radioactive rays – you can use them all to make food safer for longer periods of time by reducing or eliminating damage from exposure to air or organisms (microbes) that live on food.

Used correctly, each method can dramatically lengthen food's shelf life. The down side? Nothing's perfect, so you still have to monitor your food to make sure that the preservation treatment has, well, preserved it. The following pages tell you how.

Cold comfort: Chilling and freezing

Keeping food cold, sometimes very cold, slows or suspends the activity of microbes bent on digesting your food before you do.

Unlike heat, which actually kills many of the microbes, chilling or freezing food doesn't kill the microbes; instead, it merely puts them on the sidelines for a while. *Mould* or *bacteria spores* (which in essence are hibernating mould or bacteria organisms), for example, lie dormant inside frozen food like hibernating squirrels. When you thaw the food, the mould or bacteria springs back into action just like the squirrels in the spring, so you need to eat or cook the food as soon as possible.

Following the chilling and freezing rules

How long food stays safe in the refrigerator or freezer varies from food to food. If food is fresh before you chill or freeze it, and the equipment maintains a constant temperature (–18 degrees centigrade for freezers and no more than 5 degrees centigrade for refrigerators) food lasts longer. If the food is stale to start with, or your fridge or freezer is playing up, food spoils more quickly. Use your common sense. If food seems in any way questionable, throw it out without tasting it.

To help stop bacteria from growing in food, follow these guidelines:

- ✔ When you're preparing food, keep it out of the fridge for the shortest time possible.

- ✔ If you've made a dish you don't want to eat straight away, keep it in the fridge until you're ready to eat it.

- ✔ When the label says 'keep refrigerated', do what it says!

✔ Leave buffet or party food in the fridge until the very last minute before serving, and keep it out for no longer than two hours.

✔ Cool leftovers in less than 90 minutes, store them in the fridge and eat them up or throw them out within two days.

Follow these guidelines for storing meat:

✔ Store raw meat and poultry in clean, sealed containers on the bottom shelf of the fridge, so they can't touch or drip onto other food.

✔ Follow any storage instructions on the label and never eat meat after the use-by date.

✔ Cool cooked meat as quickly as possible and then put it in the fridge or freezer. Remember to keep cooked meat separate from raw meat.

Freezing meat is fine providing you stick to the rules:

✔ Freeze meat before the use-by date.

✔ Follow any freezing or thawing instructions on the label.

✔ Defrost the meat in a microwave if you intend to cook it as soon as it's defrosted, otherwise thaw the meat in the fridge so that it doesn't get too warm. Again, stick the meat in a container with raised sides at the bottom of the fridge so it doesn't drip onto any other food.

✔ Use the meat within two days of defrosting – it goes off in the same way as fresh meat. (if you don't use it within 2 days throw it away – you can't refreeze it).

✔ Cook frozen meat until piping hot all the way through.

Lots of liquid can come out of thawing meat. This liquid can spread bacteria to any food or surfaces it touches. Always thoroughly clean plates, surfaces and hands after touching raw or thawing meat, to stop bacteria from spreading.

If you defrost raw meat and then cook it thoroughly, eat it and don't be tempted to reheat or refreeze – if you can't eat it, chuck it!

How freezing affects the texture of food

When food freezes, the water inside each cell forms tiny crystals that can tear cell walls. When the food thaws, the liquid inside the cell leaks out, leaving thawed food drier than fresh food.

Burning when freezing

Freezer burn is a harmless, dry brownish spot left when moisture evaporates from the surface of frozen food.

To prevent freezer burn, wrap food securely in freezer paper or aluminum foil and put the item in a plastic bag. The more air you keep out, the fewer brown spots will develop.

You can't restore the crispness of vegetables, such as carrots, that get their crunch from stiff, high-fibre cell walls. After ice crystals puncture the walls, the vegetable turns mushy. The solution? Remove carrots and other crunchies, such as cabbage, before freezing the stew.

Refreezing frozen food

The official word is that partially defrosted food should never be re-frozen because the food defrosts unevenly, leaving pockets of defrosted food where bacteria have had the opportunity to multiply. No matter how tempting it might seem to save wasting food it's really not worth the risk– when in doubt, throw it out!

Canned food: Keeping out contaminants

Food is canned by heating what goes into the container and then sealing the container to keep out air and microbes. Like cooked food, canned, or tinned, food is subject to changes in appearance and nutritional content. (See the sidebar 'The essence of canned food'.) Heating food often changes its colour and texture. It also destroys some vitamin C. But tinning effectively destroys a variety of pathogens, and it deactivates those troublesome enzymes.

A modern variation on tinning is the sealed plastic or aluminium bag known as the *retort pouch*. No, it's nothing to do with the food being unable to answer back. Food sealed in the pouch is heated, but for a shorter period than that required for tinning. As a result, the pouch method does a better job at preserving flavour, appearance and heat-sensitive vitamin C.

The essence of canned food

The technique of food canning or tinning was discovered (depending on your source) either in 1809 or 1810 by Nicholas Appert, a Frenchman who noted that if he sealed food in a container while it was heating, the food stayed edible longer – much longer – than fresh food.

According to Harold McGee, author of *On Food and Cooking*, a wonderful guide to food technology, a tin of 114-year-old tinned meat was once eaten without making anyone sick. To be fair, we must note that nobody cried 'Oh, wow, this is delicious' either. Not one to try at home.

The sealed tin or pouch also protects food from deterioration caused by light or air, so the seal must remain intact. When the seal is broken, air seeps into the tin or pouch, spoiling the food.

A more serious hazard associated with canned food is *botulism*, a potentially fatal form of food poisoning caused by the failure to heat the food to high enough temperatures or for a long enough time to kill all Clostridium botulinum bacteria. *C. botulinum* is an *anaerobic* (*an* = without; *aerobic* = air) organism that thrives in the absence of oxygen, a condition nicely fulfilled by a sealed tin. Botulinum spores not destroyed by high heat during the tinning process may produce a toxin that can kill by paralysing your heart muscles and the muscles that enable you to breathe.

To avoid tinned food contaminated with botulinum organisms, do not buy, store or use any tin that is:

✔ Swollen, which indicates that bacteria are growing inside and producing gas

✔ Damaged, rusted or deeply dented along the seam, because a break in the tin permits air to enter and causes the food to spoil

Chapter 20

Weird Science: Examining Food Additives

*T*his chapter is about food additives, the substances added to food to help in the manufacturing process; to enhance the appearance, flavour or texture of the food; or to keep it safe and fresh for longer. In the United Kingdom additives are only permitted for use in food if they're deemed both safe and necessary, and they must always be displayed on the ingredients list. All approved additives are given a number: those approved for use throughout the European Union are also prefixed by a letter – yes, you've guessed it – the so-called E numbers!

Food additives can be natural or synthetic. For example, did you know that vitamin C is E number E300? E300 is added to foods as a natural preservative. *Butylated hydroxianisole* (BHA or E320) and *butylated hydroxytoluene* (BHT or E321) are synthetic chemical preservatives often used instead. The only difference between them is that the first occurs naturally in food and the second two are created in a laboratory.

Many people think that natural additives are safer than synthetic ingredients because *synthetic* seems synonymous with *chemical*. Besides, manmade additives often have names no one can pronounce, much less translate, which makes them even more user unfriendly. In fact, everything in the world is made of chemicals – your body, the air you breathe, the paper this book is printed on, as well as all food and drink. The trick with food additives is simply to use safe chemicals, whether natural or synthetic, at safe levels.

Exploring the Purpose of Food Additives

The list of common food additives includes:

- ✔ Colouring agents
- ✔ Flavours and flavour enhancers
- ✔ Preservatives
- ✔ Stabilizers, emulsifiers and thickeners

We delve deeper into these additives in the following sections.

Colours and flavours

Colouring and flavouring agents make food look and taste better.

Colours

Sometimes natural colour is lost when foods are processed. Manufacturers add colourings to foods to restore them to their original glory. Tinned peas contain a colour called green S (E142) – without it they'd be a grey mush and no one would fancy them much!

Natural colourings include *beta carotene* (E160), the natural yellow pigment in many fruits and vegetables, which can make margarine (which is naturally white) look like creamy yellow butter. Other natural colouring agents are *curcumin* (E100), a yellow pigment from turmeric roots, added to baked goods, sauces and soups; *chlorophyll* (E140), the green pigment from green plants; and *cochineal* (E120), a red pigment from a female beetle. Artificial colours include coal tar dyes such as sunset yellow (E110) found in some soft drinks, puddings and confectionery.

Flavours

Artificial flavours reproduce the taste and smell of natural flavours. For example, in processed foods artificial lemon flavouring works just as well as fresh lemon juice. Flavours don't have E numbers; nor are they always listed separately by name on a food label.

Flavour *enhancers* are a slightly different kettle of fish. They intensify a food's natural flavour instead of adding a new one. The best-known flavour enhancer is *monosodium glutamate* (MSG; E621), which is used in soups, sauces and ready meals.

MSG may trigger headaches and other symptoms in a small number of people who are sensitive to the seasoning.

Preservatives

Additives from this group are used to keep food safe and tasty for longer. Food spoils in many different and unpleasant ways. Milk turns sour. Bread goes mouldy. Meat and poultry rot. Vegetables lose moisture and wilt. Fats go rancid.

The first three kinds of spoilage are caused by *microbes* (bacteria, mould and yeasts). The last two happen when food is exposed to oxygen (air).

All preservative techniques – cooking, chilling, canning, freezing, drying, pickling and salting – prevent spoilage either by slowing the growth of the organisms that live on the food or by protecting the food from the effects of oxygen. Chemical preservatives do essentially the same thing:

- **Antimicrobials** are preservatives that protect food by slowing the growth of bacteria, moulds and yeasts. For example, sodium nitrite (E250) is very effective at preventing the growth of the bacteria that can cause *botulism* (poisoning from tinned food).

- **Antioxidants** are preservatives that protect the fats in food by preventing them from combining with oxygen (air). One of the most widely used is vitamin C (E300).

Table 18-1 is a representative list of some common preservatives and the foods in which they're found.

Table 18-1	Preservatives in Food
Preservative	*Found In . . .*
Ascorbic acid (vitamin C) and ascorbates (E300–E304)	Cured meats, sausages, luncheon meats
Benzioc acid and benzoates (E210–E219)	Soft drinks, ice cream, confectionery, salad dressings, sauces
Butylated hydroxyanisole (BHA) (E320) and butylated hydroxytoluene (BHT) (E321)	Crisps and other fried foods
Nitrites and nitrates (E249–E252)	Bacon, ham, other cured meats
Propionic acid and propionates (E280–E283)	Baked goods, bread, processed cheese
Sulphur dioxide and sulphites (E220–E228)	Processed fruit and vegetables including dried varieties, soft drinks, sausages
Tocopherols (vitamin E) (E306–E309)	Fats, oils, sausages

Some other additives in food

Food chemists use a variety of other natural and chemical additives to improve the texture and appearance of food:

- ✔ **Emulsifiers** such as lecithin (E322) keep ingredients like oil and water together.
- ✔ **Thickeners and gelling agents** are natural gums and starches, such as pectin (E440) and locust bean gum (E410), that add body to foods.
- ✔ **Stabilisers,** such as the alginates derived from seaweed (E400–E405), stop ingredients separating and make desserts such as ice cream and instant desserts feel smoother and creamier in your mouth.

Although many of these additives are derived from foods, their real benefit is for technical rather than nutritional reasons.

Determining the Safety of Food Additives

In the United Kingdom food additives are carefully regulated and assessed for safety. Putting anything into food that will knowingly injure the health of the consumer is illegal. All E numbers are tested, approved and reviewed in the light of new research. The safety of any chemical approved for use as a food additive is based on whether it's:

- ✔ Toxic
- ✔ Carcinogenic
- ✔ Allergenic

Toxins

A *toxin* is a poison. All E numbers undergo thorough toxicology testing and are considered nontoxic in the amounts that are permitted in food.

Carcinogens

A *carcinogen* is a substance that causes cancer. Reports sometimes claim that certain food additives have been identified as cancer-causing agents. For example, we know that nitrates and nitrites are effective preservatives. However, when they reach your stomach, nitrates and nitrites react with natural ammonia compounds called *amines* to form *nitrosamines*. These substances are known to cause cancer in animals, albeit at levels much higher than those used to preserve food. Levels of these nitrates and nitrites used as preservatives are now considerably reduced. A vitamin C compound (sodium ascorbate) or vitamin E compound (tocopherol) is also added to cured meats to prevent the formation of nitrosamines while boosting the antimicrobial powers of the nitrates and nitrites.

Some colourings are banned from use in the United Kingdom and the rest of the European Union even though the risk of cancer associated with them may be quite small. An example to hit the headlines recently is the Sudan dyes, red colouring agents and suspected carcinogens, illegally added to some brands of chilli powder imported to the United Kingdom. Foods contaminated with these colours were withdrawn or recalled from the shelves to avoid unnecessary consumer exposure, however small the risk.

Allergens

Allergens are substances that trigger allergic reactions in susceptible individuals (see Chapter 21 for loads more on food and allergies). Some allergens, such as peanuts, can provoke fatal allergic reactions in those with a peanut allergy.

Some additives may also provoke an allergic reaction in people who have asthma or are prone to other allergic reactions such as nettle rash (urticaria) or eczema.

An example of additive allergens comes from a group of preservatives found in soft drinks, meat products and dried fruits. They include sulphur dioxide (E220) and the other sulphites (E221–E228). These additives are safe for most people but not for all. People with asthma can react to them, developing breathing problems simply by inhaling fumes from sulphite-treated foods. Even infinitesimally small amounts may trigger a serious allergic reaction in some asthmatics. Preservatives known as *benzoates* (E 210–E219) also found in soft drinks can also exacerbate asthma and eczema in some children.

Going Hyper: Food Additives and Children

The question of whether food additives can cause behavioural problems in children remains an area of uncertainty and concern for nutritionists and the public alike. Although all approved additives are tested for safety individually, no one really knows the long-term effects of using them in combination.

Hyperactivity (or *Attention Deficit Hyperactivity Disorder*, ADHD) is a pattern of behaviour including reduced attention span and problems concentrating, which can affect learning and cause impulsive behaviour at home and at school. ADHD is difficult to define and diagnose, but experts believe that 2 to 5 per cent of children in the United Kingdom show signs of it.

The following seven additives, which you find in a range of foods including soft drinks, sweets and cake, have been implicated in hyperactive behaviour in children:

- ✔ The artificial colours tartrazine (E102), quinoline yellow (E104), sunset yellow (E110), carmoisine (E122), ponceau 4R (E124) and allura red (E129)
- ✔ The preservative sodium benzoate (E211)

A lot of the evidence linking additives and ADHD was originally anecdotal, but in 2007 researchers at Southampton University carried out a government-funded trial in which they put these seven additives into nonalcoholic cocktails and gave them to 300 young children daily for six weeks. They observed the children's behaviour compared with a placebo drink free from additives. Some links were found between the additives and activity and attention in some children However, it's important to note additives aren't the only cause of ADHD and factors such as genetics, prematurity and upbringing also play a role. Manufacturers and retailers have done little about the use of benzoate preservatives and the use of colours in children's medicine. But some in the UK have bowed to pressure from the Food Standards Agency as well as the consumer for a voluntary ban and replaced artificial colours in their processed foods, or at least those aimed at children. Some retailers have gone as far as replacing all artificial colours as well as flavours in their whole ranges, or wherever a food or drink contains additives, clearly displaying on the label either the E number or the name.

From July 2010 a European Union-wide mandatory warning label has been introduced on any food and drink (except drinks with more than 1.2% alcohol) that contains any of the six colours. The label must carry the warning 'may have an adverse effect on activity and attention in children'. Food and drink produced before 20 July 2010 can continue to be marketed, so it may take time for newly labelled products to appear on some store shelves. If you buy any foods that are sold without packaging you will need to check with the person selling the product or with the manufacturer.

So now you do have more choice if you want to avoid certain artificial additives.

Gene Cuisine: Debating Genetically Modified Food

Genetic modification (GM) is a type of biotechnology that allows scientists to take individual genes or small groups of genes carrying particular characteristics from a plant or animal, alter the genes, copy them and then insert them into another plant, animal or microorganism to change it in a specific way. For example, this process can make a plant resistant to a particular disease or to a particular herbicide that might usefully kill all the weeds around it. This method while similar to traditional methods of selective breeding, used for thousands of years by farmers, differs as unusually it allows scientists to exchange genes across completely unrelated species and even between plants and animals.

Intense debate rages over GM foods. Some people believe that gene technology is Frankenstein food, tampering with nature and playing God. Others believe that patenting GM processes could be the answer to world problems of over-population, resource depletion and climate change.

Looking at the pros and cons

Here are some of the key arguments for and against GM foods.

Benefits include:

- GM crops that are pest resistant and give higher yields could provide enough food for the ever-expanding global population.
- Plants can be modified to be more nutritious or have greater health properties.
- GM plants can be developed to survive extreme conditions like drought that may occur more often in the future given climate change fears.
- Pesticides and herbicides may be used less intensively on GM crops.
- GM foods can be developed to provide health benefits like edible vaccines.
- GM foods can provide cheaper, better quality, tastier food.
- GM foods appear safe. Your digestive system is designed to break down GM foods, including their genes, in exactly the same way as it does with non-GM foods. No reported cases of illness or disease caused by eating a food that's been genetically modified exist.

Possible risks include:

- ✔ We don't know enough about what happens to genes inserted into GM crops.

- ✔ Growing GM crops on a large scale may have implications for *biodiversity* – the balance of nature and the environment including the production of resistant 'superweeds'.

- ✔ Genes from GM crops can transfer to non-GM crops and other plants nearby.

- ✔ Using antibiotic resistant genes as GM markers or measures could add to antibiotic resistance in humans.

- ✔ Toxins or allergens may be increased, transferred or produced as a result of GM.

- ✔ Animals may be exploited. Research is looking into ways to make animals grow faster and bigger and to assist human transplantation.

The European Food Safety Authority assesses any GM foods or those containing GM ingredients on sale in the EU for safety including toxicity, nutrition and possible allergenicity. Details of every application are available on their website (www.efsa.europa.eu).

To date, no genetically modified organisms (GMOs) themselves have been approved for use in the EU but foods containing GM derived ingredients such as flour, oils and glucose syrup derived from GM crops such as rapeseed, maize and soy are available. However, scientists predict that the first GM crops – probably GM potatoes resistant to common diseases – will be grown in the UK by 2015 if the results of current research trials prove successful.

The Food Standards Agency oversees the use of GM ingredients in the UK and says it is 'satisfied that the current safety assessment procedures for GM foods are sufficiently rigorous to ensure that GM foods are as safe as non-GM counterparts, and pose no risk to the consumer' (http://www.food.gov.uk/gmfoods/gm/).

Knowing what's GM and what's not

In April 2004 UK legislation regarding the labelling of GM foods came into force. Unlike in the USA and Canada, in the UK labels must now indicate that a food contains or consists of GMOs, or contains ingredients derived from GMOs. If the food is sold loose, this information must be displayed next to the food. You might think this makes it simple to decide whether or not to buy GM foods, but as with most legislation where food labelling is concerned, unfortunately it's more complicated than that! Products produced with GM *technology* (not with modified genes specifically) don't have to be labelled. The label doesn't have to state that the cheese you buy was made with enzymes derived from GM bacteria or that a food contains small (less than 1 per cent) amounts of accidentally present but approved GM ingredients, such as from cross-contamination of crops. Also, your meat, milk and eggs might be from animals fed on GM crops, but the label won't always tell you. Regardless of where you sit on the GM debate, you may well be eating food that has some sort of GM association without even realising it!

In 2010 a survey showed that 72 per cent of UK families wanted their food guaranteed non-GM and were prepared to pay extra to ensure this. However, fewer than 40 per cent were aware that over a million tonnes of GM crops are imported each year to the UK to feed non-organic livestock that produce pork, bacon, chicken, eggs, milk, cheese and other dairy products they buy. At present some retailers will sell only meat, poultry and eggs reared on non-GM feed, but they're under constant financial pressure to change this.

The label 'GM-free' may be present on products, but until this becomes a legally defined and recognised term, it's potentially misleading to the consumer. Meanwhile, GM crops and ingredients continue to be banned under organic standards.

Considering Organic Options

Organic food is framed and produced under very strict rules that control the methods the producer can and can't use. These rules go far beyond a ban on the use of GM ingredients and certain food additives, and aim to protect natural resources, wildlife and livestock welfare. The rules include severe restrictions on the use of pesticides, and prohibit artificial chemical fertilizers, animal cruelty and antibiotics and other drugs in healthy animals. They ensure for farming techniques that promote free-range lifestyles, careful animal husbandry, natural pest control, natural fertilisers and crop rotation.

Organic farming methods are often labour intensive, meaning organic foods come at a price. Despite this, the UK has seen massive growth in organic food sales in recent years, with the organic market worth £1.8 billion in 2009.

Fans of organic food say it's better for your health than non-organic food. But just how true is this? Not all experts are in agreement. They maintain that all food, whether organic or not, must meet the same legal safety standards. Pesticides used in conventional foods are approved for use and residue levels are monitored and regulated to ensure they don't exceed safe levels. The same applies to food additives. And the differences between natural and artificial fertilisers are very blurred.

Recently, independent scientific research commissioned from the Food Standards Agency looked at all the studies carried out comparing the differences in nutrition and health effects between organic and conventional foods. The research focused on the findings on over 50 scientifically well-designed studies and came to the conclusion that, in terms of nutrition, no major differences existed between organic and non-organic. Nutrient levels were found to vary more according to the local climate, soil, animal feedstuff used and storage conditions of the food. Some organic fruit and vegetables had slightly higher levels of some plant antioxidants and organic milk was slightly higher in short chain omega-3 fatty acids, but neither of these factors were considered to have a major impact in practice. Other than one study that suggested organic dairy products were more beneficial in reducing the risk of eczema in infants, no studies showed a benefit of organic foods on health.

Organic certification covers the wider issues of environment and animal welfare as well as food choice. Although it's hard at present to justify the extra cost of organic foods on health grounds alone, other benefits of organic farming may sway your decision between organic and non-organic foods.

Seeking Sustainability

Closely linked to issues of additives, GM foods and organic farming is the idea of the sustainability of food and agricultural systems.

Sustainability is about balancing food production with the use of natural resources including water, air, land and forest, and avoiding harm to people and animals including wildlife and fish stocks. In recent years sustainability issues have become as important as nutrition, taste and cost for many consumers in influencing food choices. UK food production and consumption accounts for up to 25 per cent of the country's carbon footprint, contributing significantly to greenhouse gas emissions. Sustainable food practices are likely to become even more important as the UK moves towards achieving its target of reducing its carbon footprint by 80 per cent by 2050. Food provides an opportunity to vote three times a day, using your consumer power, on where you want the country to go in terms of sustainable practices.

Probing provenance

A key concern for food these days is *provenance* – where and how food is produced and how far it's travelled to get to your plate (food miles). These factors can influence animal welfare, local growers and climate change – for example, via the greenhouse gas emissions involved in production, packaging and transport.

The Institute of Grocery Distributors (IGD) recently published figures showing that 17 per cent of UK shoppers said they'd increased the amount of local food they bought in the preceding six months from sources such as local growers, co-operatives and farmers' markets. Buying British is becoming more popular. And according to a Which? poll, over 70 per cent of consumers in the UK want more information about where their food comes from. Some retailers have started to do voluntary *county of origin labeling* (COOL) on foods, but this is unlikely to become mandatory in the near future.

Where foods can't be sourced locally, fair trade certification is useful if you want to ensure the producers in developing countries get a fair price for their crops such as sugar, tea, coffee, cocoa beans or bananas. The IGD survey showed shoppers are very willing to buy sustainable and ethical foods even in credit crunch times if they perceive see genuine benefits.

Other certification schemes can also help you know when the food you're buying is making a difference. In the IGD survey consumers reported a 22 per cent increase in products promising high animal welfare standards. Organic certification and other markers, such as the RSPCA Freedom Food Standard, ensure high levels of animal welfare in production, including fish farming. For many, the reduction in population of a wide range of fish species by industrialised over-fishing during the past 50 years is a major concern. If this worries you, look for independent assessment such as the Marine Stewardship Council (MSC) 'blue tick' on products, which ensures the fish are caught with minimal impact on stocks, eco systems and the environment. Both the nutrition and palatability of farmed and responsibly managed oily fish has been shown to be at least as good as that of wild fish.

Eating less meat?

Many environmentalists believe that the best way to operate a sustainable food system is for us to eat less meat. In 2006 the United Nations reported that worldwide the livestock industry alone is responsible for more climate changing greenhouse gas emissions than transport. It also contributes to environmental destruction and resource depletion – it may surprise you, but it takes 10 kilograms of grain and 18,000 litres of water to produce just 1 kilogram of beef.

However, to balance the argument we need to acknowledge that in recent years the livestock industry has made good progress in reducing its carbon footprint. Meat is a good source of nutrients such as iron and zinc. But reductions in portion sizes or frequency of consumption (or both!) and eating more plant foods such as pulses, beans, nuts and fortified or wholegrain cereals would shift most people's diets more closely in line with guidance from most national dietary guidelines, including the UK's Eatwell plate (see Chapter 14).

Evidence suggests that some people are taking the idea of cutting back on meat on board already. Government figures released in 2010 suggest that people in the UK are eating around 5 per cent less meat than five years ago. For more suggestions, go to www.meatfreemondays.co.uk.

Reducing food waste

Forty per cent of food and drink purchases – an amazing 8.3 million tons of food and drink – go into the bin or landfill each year.

Here are some tips to cut down on wastage:

✔ Plan your meals and shop with a list so you don't buy or cook more than you need. Look in cupboards and the refrigerator before you go shopping and don't shop on an empty stomach!

✔ Look back at your cash register receipts at the end of the week to see what, if anything, was wasted, and take a lesson for your next shop.

✔ Buy perishable foods in smaller amounts more frequently or loose packed in the amount you need and store them sensibly. Buy some fruit and vegetables when ripe and some to ripen and eat later. Storing fruit in the refrigerator extends its life by up to two weeks. Use overripe fruit in smoothies.

✔ Use leftovers to make soups and sauces for pasta, in omelettes and stews or in a packed lunch rather than relying on pre-made sandwiches or fast food. To make them more appealing, store leftovers in waterproof and covered airtight containers.

For more ideas go to www.lovefoodhatewaste.com.

Top tips for sustainable eating

Eat foods in season wherever possible. Food grown in natural sunlight is most energy efficient. Fresh, out-of-season produce is often air freighted and accounts for a large proportion of fruit and vegetables currently eaten in the UK. A home delivery box scheme or growing some of your own fruit and veg are other great ways to eat local. For more ideas, go to www.eat seasonably.co.uk.

Try to avoid bottled where the local tap water is safe to drink, such as in the UK. Use a jug or mains filter or flavourings if necessary. If do you use bottled water choose recyclable bottles and look for more eco-friendly brands.

Part V
Food and Health

In this part . . .

Here you'll find out about some of the more unusual ways that food can impact on health. We show you fascinating research on the issue of food allergies in Chapter 21, and the relatively new but exciting field of mental health and nutrition, with the effects of food on mood in Chapter 22.

In Chapter 23 we explore foods used to treat medical conditions and look at what happens when foods and drugs fight (the technical term is *food/drug interactions*). Finally in this section we seek out the truth from the mysterious and confusing world of food supplements in Chapter 24.

Chapter 21

Food and Allergies

*A*pproximately 5 to 8 per cent of all children and only 2 per cent of adults in the UK are affected by true food allergies. Many childhood allergies seem to disappear when children grow older. If those figures seem low to you then you're not alone. Up to 20 per cent of UK adults believe they have a food allergy, but numerous well-conducted trials repeatedly show that the true figure is nearer 2 per cent.

If food allergies aren't as common as people think, why read this whole chapter on them? First, although food allergies are still relatively rare, they're more common now than they were 20 years ago and it doesn't look as if they'll be going away. Food allergies that don't disappear can trigger reactions ranging from trivial (a stuffy nose the next day) to the truly dangerous (respiratory failure). In addition, whenever you're allergic to foods, you're likely to have other allergies that are triggered by such things as dust or pollen or the family cat. So knowing which food does what really pays off. After all, forewarned is forearmed.

Finding Out More about Food Allergies

Your immune system is designed to protect your body from harmful invaders, such as bacteria. Sometimes, however, your immune system responds to substances that are normally harmless. A food allergy is just such a response – your body fighting back against specific proteins in foods. Some common allergic reactions to food include:

✔ Asthma

✔ Breathing difficulties caused by tightening (swelling) of tissues in the throat

✔ Diarrhoea, sometimes bloody

✔ Headache, migraine

✔ Hives (*urticaria*), an acutely itchy, red patch of skin with a pale fluid-filled centre

✔ Itching

✔ Loss of consciousness (anaphylactic shock)

✔ Nausea and/or vomiting

✔ Rashes

✔ Sneezing, coughing

✔ Swelling of the face, tongue, lips, eyelids, hands and feet

Investigating allergies and intolerances

Your body may respond to a food in one of two ways:

✔ **Food allergy:** Specific reaction resulting in an *immunological response* (where the body produces a dramatic reaction in the presence of an allergen) to a food that can be severe or life threatening. Symptoms can be rapid or delayed. Immediate reactions are more dangerous than delayed reactions because they involve a fast swelling of tissues. Immediate reactions may occur within seconds after eating, touching or – in some cases – even smelling the offending food.

✔ **Nonallergic food intolerance:** Reactions to food that can result from a number of causes, but that aren't a result of an immunological response. Reactions are usually as a result of histamine release, an interaction with medication, or enzyme deficiency, and often occur after eating a relatively large amount of a specific food. Reactions may occur as long as 24 to 48 hours after you're exposed to the offending food, and the reaction is likely to be much milder, perhaps a slight nasal congestion caused by swollen tissues.

These two responses are often confused. People with a food intolerance can often cope very well with a small amount of a food that in larger quantities would cause them a problem; they can also often manage a problem food when it's been cooked or processed in some way. For instance, people who are intolerant to milk may be fine with yogurt, or people who react to tomatoes can cope with tomatoes cooked in a sauce with pasta.

Allergy lingo

This allergy glossary can help you understand what's going on with allergies.

Allergen: Any substance that sets off an allergic reaction (see 'antigen' in this sidebar)

Anaphylaxis: A potentially life-threatening allergic reaction that involves many body systems

Antibody: A substance in your blood that reacts to a specific antigen

Antigen: A substance that stimulates an immune response; an allergen is an antigen

Basophil: A type of white blood cell that carries IgE (see 'IgE') and releases histamine

ELISA: Short for *enzyme-linked immunosorbent assay*, a test used to determine the presence of antibodies in your blood, including antibodies to specific allergens

Histamine: The substance released by the immune system that provides symptoms or an allergic reaction such as itching and swelling

IgE: An abbreviation for *immunoglobulin E*, the antibody that reacts to allergens

Intolerance: A nonallergic adverse reaction to food

Mast cell: A cell in body tissue that releases histamine

RAST: An abbreviation for *radioallergosorbent test*, a blood test used to determine whether you are allergic to certain foods

Urticaria: The medical name for hives, an itchy rash

Source: American Academy of Allergy & Immunology, International Food Information Council Foundation, 'Understanding Food Allergy' (April 1995)

Avoiding any food unnecessarily can increase your risk of nutritional deficiency. If you have a true food allergy you can't be as flexible. For example, if you're allergic to peanuts you need to steer clear of them in any amount or any form.

Head for Accident and Emergency or call 999 immediately if you – or a friend or relative – show any signs of an allergic reaction that affects breathing.

Understanding how a reaction occurs

When you eat a food containing a protein to which you're allergic (the *allergen*), your immune system releases antibodies (*IgE*) that recognise that specific allergen. The antibodies circulate through your body on white blood cells (*basophils*) that pass into all your body tissues, where they bind to immune system cells, called *mast cells*.

Basophils and mast cells produce, store and release histamine, which causes the symptoms – itching, swelling, hives – associated with allergic reactions. (That's why some allergy pills are called antihistamines.) When the antibodies carried by the basophils and mast cells come in contact with food allergens, boom! You have an allergic reaction.

Food intolerances can cause similar symptoms to food allergies, but the main difference is that the immune system isn't involved. The body releases histamine as a result after mast cells are activated directly without the need for antibodies.

Often the most important difference between a food allergy and a food intolerance is the way in which you can diagnose each. A food allergy is far easier to diagnose than an intolerance because you can measure the amount of immunological activity in the blood. With an intolerance you have nothing to measure, but more about that later in the section.

It's all in the family: Inheriting food allergies

You can inherit a tendency towards allergies, although not the particular allergy itself. If one of your parents has a food allergy, your risk of having the same problem is two times higher than if neither of your parents is allergic to any foods. If both your mother and your father have food allergies, your risk is four times higher. In one truly unusual incident, a transplant patient got a severe peanut allergy along with a new liver from a peanut-sensitive donor.

Being aware of the dangers of food allergies

Food allergies can be dangerous. Although most allergic reactions are unpleasant but essentially mild, a small number of people die every year from allergic reactions to food. These people have suffered *anaphylaxis*, a rare but potentially fatal condition in which many different parts of the body react to an allergen in food (or some other allergen), creating a cascade of effects beginning with sudden, severe itching, and moving on to swelling of the tissue in the air passages that can lead to breathing difficulties, falling blood pressure, unconsciousness and death. Many people with true food allergies carry an EPIPEN that contains the hormone epinephrine. If you inject epiniphrine early during an anaphylactic attack it can be life saving.

Considering Foods Most Likely to Cause Allergic Reactions

Here's something to chew on: more than 90 per cent of all allergic reactions to foods are caused by just a small selection of food:

- celery
- cereals containing gluten (including wheat, rye, barley and oats)
- crustaceans (including crabs and prawns)
- eggs
- fish
- lupin
- milk
- molluscs (such as mussels and oysters)
- mustard
- nuts (including Brazil nuts, hazelnuts, almonds and walnuts)
- peanuts (groundnuts or monkey nuts)
- sesame seeds
- soya
- sulphur dioxide or sulphites

The most common foods to cause intolerance reactions are:

- Cheese (especially matured cheese)
- Chocolate
- Coffee and foods and drinks containing caffeine
- Fermented foods such as blue cheese and fermented soya products
- Fruits (especially citrus fruits, avocado and banana)
- Red wine
- Wheat and gluten
- Lactose
- Yeast extracts

The allergenic food most likely to make headlines seems to be peanuts. People allergic to peanuts may break out in hives just from touching a peanut or peanut butter, and may suffer a potentially fatal reaction after simply tasting chocolate made in a factory where the chocolate had touched machinery that had previously touched peanuts.

For information about potentially allergenic food additives, see Chapter 18.

Testing, Testing: Identifying Food Allergies

Whenever you sprout hives or your skin itches or your eyelids, lips and tongue begin to swell right after you've eaten a particular food, that's a clear sign of a food allergy. Some allergic reactions, however, occur in milder forms, many hours after you've eaten. To identify the culprit, your doctor may suggest an *elimination diet*. This regimen removes from your diet foods known to cause allergic reactions in many people. Then, one at a time, you add the foods back. If you react to one, bingo! That's a clue to what triggers your immune response. An elimination diet is hard work; it takes a lot of effort and requires the support and guidance of a registered dietitian to avoid nutrient deficiency. Exclusion diets should only be followed for a short period of time.

A dietitian can also help you find which ingredients appear in the most unlikely foods – for instance, would you ever guess that many brands of ice cream contain wheat? To be absolutely certain of an allergy, your dietitian may challenge your immune system by introducing foods in a form (maybe a capsule) that you can't identify as a specific food. Doing so rules out any possibility that your reaction has been triggered by emotional stimuli – that is, seeing, tasting or smelling the food.

Two more sophisticated tests – *ELISA* (enzyme-linked immunosorbent assay) and *RAST* (radioallergosorbent test) — can identify antibodies to specific allergens in your blood. But these two tests are rarely required.

Because food allergies and intolerance are so topical, a plethora of therapists with dubious qualifications offer tests with no scientific basis, relieving you of your cash in exchange for an inaccurate diagnosis. Often these tests suggest that you avoid a whole range of foods unnecessarily, putting you at risk of nutritional deficiency. If in doubt, check out the range of reliable and unreliable tests on the Allergy UK Web site: www.allergyuk.org.

Coping with Food Allergies and Intolerances

After you know that you're allergic or intolerant to a food, the best way to avoid a reaction is to avoid the food. Unfortunately, that task may be harder than it sounds.

Some allergens are hidden ingredients in dishes made with other foods. For example, people allergic to peanuts have suffered serious allergic reactions after eating chocolate containing traces of nuts. Another problem with food allergies is that you may not even have to eat the food to suffer an allergic reaction. People who react to seafood – fish and shellfish – are known to have developed respiratory problems after simply inhaling the vapours or steam produced by cooking the fish.

If you're someone with a potentially life-threatening allergy to food (or another allergen, such as wasp venom), your doctor may suggest that you carry a syringe prefilled with *epinephrine*, a drug that counteracts the reactions (an EPIPEN). You may also want to wear a medic alert bracelet that identifies you as a person with a serious allergic problem. The food industry is also legally bound to include labels on pre-packed foods containing some of the most common allergens – especially nuts and peanuts.

Chapter 22

Food and the Brain

. . . .

In This Chapter

▶ Discovering the effect of different nutrients on the brain

▶ Understanding how food affects your mood

▶ Exploring the influence of diet on behaviour, thinking and learning

. . . .

Draw the curtains, turn down the lights and put on some soft music. Are you sitting comfortably? We're going to talk about how what you eat can affect how you feel, learn, think and behave.

You're already pretty familiar with how food affects your body, but it might come as a surprise that what you eat and drink affects your mind as well. A few fascinating factoids might help you see why:

✔ Your brains account for only 2 per cent of your body weight, but you use over 50 per cent of your blood sugar supply at rest – that's a lot of activity that needs regular refueling to keep it functioning.

✔ Fat makes up an amazing 60 per cent of the dry weight of the brain, concentrated mainly in the brain cell membranes, and all this has to come from your diet.

✔ All the messages passing back and forth between brain cells use up vitamins and minerals, and these need to come from your food.

This really is food for thought!

This chapter discusses some of the common, naturally occurring chemicals in food that can alter your mood, your behaviour and even your ability to learn. Sit back, open a box of chocolates, pour a glass of wine or brew up the espresso – and enjoy!

How Chemicals Alter Mood

A *mood* is a feeling, an internal emotional state that can affect how you see the world. For example:

- ✔ If your football team wins the Premiership your happiness might last for days, making you feel so mellow that you simply shrug off minor annoyances such as getting a parking ticket when you stop to pick up a newspaper.

- ✔ If you feel sad because the project you spent six months setting up didn't work out, your disappointment can linger long enough to make your work seem temporarily unrewarding or your favourite television programme boring.

Most of the time, after a few ups and downs, your mood swings back to centre fairly soon. You come down from your high or recover from your disappointment, and life resumes its normal pace – some good news here, some bad news there, but all in all, a relatively level field.

Occasionally, however, your mood can go out of control. About 25 per cent of people (women more often than men) experience some form of mood disturbance during their lifetime. The most common mood disorder in the United Kingdom is *clinical depression*: prolonged and intense sadness, loss of interest and enjoyment in daily life and reduced activity levels.

Scientists have identified naturally occurring brain chemicals that affect mood and play a role in mood disorders. Your body makes a group of substances called *neurotransmitters*, chemicals in the brain that enable the brain cells to send messages. Neurotransmitters are made from the amino acids you get from protein in your diet. Three important neurotransmitters are:

- ✔ Dopamine and noradrenaline, the so-called 'motivational' neurotransmitters that make you feel alert and energised, which are made from the amino acid phenylalanine.

- ✔ Serotonin, the 'feel good' neurotransmitter that makes you feel happy and calm, which is made from the amino acid tryptophan.

Some forms of clinical depression appear to be malfunctions of the body's ability to make and use these chemicals. Drugs known as *antidepressants* adjust mood by increasing the availability of these neurotransmitters to your brain.

Examples of antidepressants include: tricyclic antidepressants (TCAs) such as amitriptyline; selective serotonin reuptake inhibitors (SSRIs) such as fluoxetine (Prozac) and paroxetine (Seroxat); and monoamine oxidase inhibitors (MAOIs).

The food you eat can also affect how you feel.

How Food Affects Mood

Good morning! Time to wake up, roll out of bed and crawl into the kitchen for that cup of coffee.

Good day at work? No? Time for a gin and tonic or a glass of wine to soothe away the tensions of the day.

Good grief! Your lover has left. Time for chocolate, lots of it, to soothe the pain.

Good night! Time for milk and biscuits to ease your way to sleep.

For centuries, millions of people have used foods in these situations, secure in the knowledge that each food will work its mood magic. Today, modern science knows some of the reasons foods affect mood. Having discovered that emotions are linked to the production or use of certain brain chemicals, nutritionists have identified the natural chemicals in food that change the way you feel.

The following sections describe the chemicals in food that are most commonly known to affect mood.

Alcohol

Alcohol is probably the world's most widely used natural sedative.

Contrary to common belief, alcohol is a depressant, not a stimulant. If you feel exuberant after one drink, the reason isn't because the alcohol is speeding up your brain. It's because alcohol relaxes your *controls*, the brain signals that normally tell you not to put a street cone on your head or take off your clothes in public.

Alcohol activates the pleasure or reward centres of the brain by triggering the release of both dopamine and serotonin. For more about alcohol's effects on virtually every body organ and system, turn to Chapter 8. Here in this chapter, it's enough to say that many people find that, taken with food and in moderation (defined as no more than two to three units a day for women and no more than three to four units a day for men), alcohol can produce a sense of well-being, relaxation and happiness.

Caffeine

Caffeine is probably the most widely consumed stimulant in the world. For some people, if they're not used to it, a large dose of caffeine can have a variety of effects. Caffeine can:

✔ Increase urine production

✔ Raise your blood pressure

✔ Speed up your heartbeat

✔ Stop you sleeping

However, doses at the levels commonly consumed have little or no ill effect, especially if you're a regular consumer.

But even in quite low doses, caffeine is mood activating – increasing alertness, especially when you're tired. This effect seems to be genuine and not simply the reversal of withdrawal effects when caffeine is withheld. Some people report higher levels of anxiety after exposure to large single doses of caffeine. However, how people react to caffeine is a highly individual affair. Some can drink several cups of normal coffee a day and experience none of the side effects; others tend to get jumpy even on quite low intakes. This may relate to body composition, weight, sex, whatever else a person's eaten recently or even a biochemical difference in the brain. Nobody really knows yet.

Either way, caffeine's effects may last anywhere from one to seven hours. Some researchers believe that children may be particularly sensitive, but insufficient research has yet been carried out to determine the effect of caffeinated fizzy drinks on children's mood. For most adults the average intake of about 300 milligrams of caffeine per day seems to be harmless. However, perhaps 10 per cent of the UK population has high intakes, usually defined as greater than about 400 milligrams per day for a woman and 500 milligrams per day for a man. Use Table 20-1 to work out your own daily level.

Endurance athletes who take caffeine before an event report that it improves alertness and performance. Although caffeine consumption is no longer banned in sport at high levels, intake s may still monitored to ensure athletes haven't abused its use. Caffeine can be detected in your urine, and drinking caffeine from a variety of sources (see Table 20-1) can easily add up.

Table 20-1	**Foods and Drinks That Give You Caffeine**
Food	*Average Caffeine Content (mg/Serving)*
Coffee, filter/percolated *(mug 200ml)	140
Coffee, instant (mug 200ml)	80–100
Coffee, espresso (single shot)	75
Coffee or tea, decaffeinated (mug 200ml)	4
Tea (mug 200ml)	75

Food	Average Caffeine Content (mg/Serving)
Cocoa (200ml)	5
Cola or diet cola drinks (330ml can)	40
Plain chocolate bar (50g)	50
Stimulant ('energy') drinks (small can) ***	80

*Source: Adapted from Food Standards Agency (http://www.food.gov.uk/news/pressreleases/2008/nov/caffeineadvice) * Some larger serving sizes of coffee from high street shop chains may contain as much as 250–400 milligrams of caffeine. ** Milk chocolate has about half the caffeine content of plain chocolate. *** Some recent arrivals on the UK market from the US are around four times this strong, giving the equivalent of two strong mugs of coffee in a 60-milliletre shot.*

Carbohydrate and protein

Proteins are composed of a series of building blocks known as amino acids. Several amino acids are important in mood. The body can convert the amino acid tryptophan to the 'feel-good' neurotransmitter serotonin, which elevates and enhances mood, acting like a natural tranquilliser. Some recent studies have shown that combining tryptophan supplements with SSRI antidepressants gives better results than taking the antidepressant on its own. Good sources of tryptophan are animal proteins such as lean meat, poultry, eggs, dairy products, seeds, pulses, beans and nuts (except peanuts, which are a poor source).

A sufficient supply of glucose from a high carbohydrate diet increases the availability of tryptophan, to the brain and increases serotonin release from nerve cells in the brain. A combination of tryptophan-rich proteins and carbohydrates is therefore thought to be especially mood enhancing. Wholegrain cereals, pasta, bread, oats, rice, fruit and pulses provide a slow supply of glucose into the bloodstream for a steady release of serotonin to help moderate and stabilise mood.

Women naturally have lower levels of serotonin in their brains than men. Recent research has shown that for some women following a low-carbohydrate weight loss diet exacerbates this increasing irritability, making their moods more unsteady. If this sounds like you try a low fat weight loss diet instead!

Using drinks to manage mood

Nutrition is important in mental function and certainly plays a role in depression. What you drink won't change your personality or alter the course of a mood disorder. But certain drinks may be able to add a little lift or a small moment of calm to your day, increase your effectiveness at certain tasks or make you more alert.

✔ One cup of coffee in the morning is a pleasant push into alertness. Seven cups of coffee a day can make you jittery and feel on edge.

✔ A glass of wine or a gin and tonic is generally a safe way to relax. Three and you may be getting *too* merry (or moody).

Less is known about the long- or short-term effects of high fat and sugar intakes. Some researchers suggest that filling up on fatty and fried foods can negatively affect your mood, making you tired and apathetic. If you eat simple sugars such as glucose or sucrose on an empty stomach, you absorb them rapidly into your blood, triggering an equally rapid increase in the secretion of *insulin*, a hormone that takes the sugars quickly back out of your blood again to to the feed the cells that can use it. The result is a rapid decrease in the amount of sugar circulating in your blood, a condition known as *hypoglycaemia* (*hypo* = low; *glycaemia* = sugar in the blood) that can make you feel temporarily jumpy and irritable rather than calm. However, when eaten on a full stomach – a dessert after a full meal – simple sugars are absorbed more slowly and may exert the calming effect usually linked to complex carbohydrates (starchy foods).

In the long term, eating lots of foods that are high in fat and sugar might also mean you're filling up on 'empty calories' and missing out on the vitamins, minerals and essential fatty acids that your brain needs.

Some studies show very positive results using a balance of carbohydrate and protein-rich foods to improve mood, but the results aren't always clear cut. Other studies have found no benefit. Not all nutritionists believe the amounts of certain nutrients in the diet (such as tryptophan) are large enough to have an effect. Why not try it for yourself? After all, a good balance of carbohydrates and protein protects against a whole range of other health problems too!

Looking to the Future: New Research into Food and Mood

The field of food and mood is growing all the time and fascinating studies hit the scientific journals almost every month. In this section we look at some of the latest research and try to sort the fact from the fiction.

Essential fatty acids

Essential fatty acids (see Chapter 5) may influence the way you feel. Ten per cent of the fat in your brain is the long chain omega-3 group of essential fatty acids that tends to be quite low in diets today. Research suggests that omega 3s may affect mood by keeping brain cell membranes flexible and fluid, and enabling neurotransmitters to signal effectively as they pass messages back and forth in the brain.

Populations with low intakes of fish, the main source of long chain omega-3 fatty acids, tend to have higher rates of many types of depression – including conditions like post-natal depression and seasonal affective disorder – than those with good intakes, such as Japan and Iceland. Researchers have found that individuals with low dietary intakes of fish (less than one to two servings a week) and also those with low blood levels of omega 3 tend to have a higher risk of depression – the lower the level, the more severe the mood disturbance. Research also found that heart attack victims who were advised to increase their omega 3 intake from oily fish (to reduce the risk of having another heart attack) seemed to have improved levels of mood over those not given this advice. Other researchers have shown that omega 3 improves the recovery of some people on antidepressants. Although more research needs to be done in this area, the existing research findings perhaps give you yet another reason to eat those oil-rich fish. You can get long chain omega 3 fatty acids from fish (aim for at laest 2 portions of fish per week one of which should be oily) or from fish body oil supplements, aiming for around 1 gram of omega 3 per day.

Selenium and folate

Both selenium and folate play a role in determining the rate of conversion of amino acids to neurotransmitters in the brain. Surveys suggest that dietary intakes of both may be marginally low in some groups of the population and that these low intakes may be linked to a higher risk of depression as well as a poorer response to treatment for depression. Whether these low levels are a result or a cause of low mood is unclear, but supplements of selenium and folic acid in some trials seemed to improve mood – in some cases, by as much as 50 per cent – in those with the lowest levels of these nutrients.

Fruit and vegetables, pulses, wholegrain or fortified cereal products and offal contain selenium and folate. Selenium also comes from meat, fish, eggs and nuts – especially brazil nuts. Ensure that you have an adequate intake, especially because these nutrients seem to play a variety of other roles in the body including possible protection against heart disease, cancer and dementia. (Head to Chapter 10 for the UK reference nutrient intakes (RNIs), because high doses are unsafe.)

Breakfast

You've no doubt heard nutritionists say that breakfast is the most important meal of the day. Well, there's a lot of truth in it, we promise! After a night's sleep you need to '*break* your *fast*' and provide your brain with fuel to get it working for you.

Low levels of blood sugar impair brain function, especially memory and concentration, and performance in other mental tasks. Low blood sugar can even affect how you deal with stressful situations. Children and older adults are especially sensitive. In a recent study children who ate breakfast scored 22 per cent higher on a word recall test than those who'd skipped breakfast.

Remembering the brain needs carbohydrate as its main fuel helps you to choose the ideal breakfast foods, like breakfast cereals and bread, that help keep your blood sugar levels in the normal range.

When you eat low glycaemic index (low GI) foods, your body absorbs and digests the carbohydrate more gradually, which helps sustain blood sugar levels for longer into the morning. Try a bowl of porridge oats or muesli with dried fruit and low-fat milk, or wholegrain toast with a poached egg or some baked beans.

Chocolate

It may come as no surprise, but a recent Mintel survey showed that consumers were willing to make certain changes to achieve a healthier diet but most were reluctant to give up chocolate, considering it a vital 'mood food' – satisfying their emotional needs when feeling low. So is there any substance to this claim? Well yes, perhaps!

Chocolate products, especially those with a high cocoa content (more than 70 per cent cocoa beans), contain a range of pleasure-enhancing ingredients. *Anandamide*, also known as 'the bliss molecule', is a neurotransmitter that mimics the euphoric and heightened sensitivity effect of cannabis in the brain. In addition, chocolate contains two chemicals that slow the breakdown of the anandamide already produced in your brain, thus intensifying its effects. (You'd need to eat at least 100 bars of chocolate at one time to get any real marijuana-like effect.)

Chocolate also contains trace levels of *phenylethylalanine* (PEA) – a stimulant similar to amphetamine that your body releases when you're in love, making you feel good all over. Low levels of PEA have been linked to depression. Chocolate also contains caffeine and another stimulant known as theobromine.

Caution! Medicine at work

Some of the mood-altering chemicals in food interact with medicines. As you may have guessed, the two most notable examples are caffeine and alcohol.

- Caffeine makes painkillers such as aspirin and acetaminophen more effective. On the other hand, many over-the-counter painkillers already contain caffeine. If you take

a pill with a strong cup of latte, you may increase your caffeine intake past the jitters stage.

- Alcohol can increase the sedative or depressant effects of some drugs, such as antihistamines and painkillers, and at the same time alter the rate at which you absorb or excrete others.

Many researchers don't believe that the amounts of these chemicals in chocolate are strong enough for a genuine mood-altering effect in the brain. Researchers found that cocoa-filled capsules containing the same chemicals as chocolate don't satisfy cravings the way chocolate does. They argue that it's only the sweetness and pleasurable feeling of chocolate in the mouth that make people feel so happy. Spoilsports!

Always ask your pharmacist about food and drug interactions when you collect a prescription and read the label on medicine very carefully.

Linking Diet, Learning and Behaviour

Recent research has explored the effects of certain nutrients in areas of the brain that control learning and behaviour. Researchers have mainly studied children and adolescents, because they are at a time of life when the brain is growing rapidly, learning and behaviour patterns are becoming established and nutrient needs are especially high. Unfortunately, childhood and adolescence are also times when diets tend to become quite poor and intakes of many vital nutrients decrease.

Using omega 3 to help with learning difficulties

Some researchers believe that omega-3 fatty acid intake has a major effect on both learning and behaviour in young people. Some evidence suggests that increased intakes of omega 3 can benefit children with learning problems such as dyslexia (difficulties with spelling and reading) and autism (characterised by impaired social and communication skills). Some children with learning problems have benefited from supplements of omega 3 and a handful of studies have shown in a scientific way that giving an omega 3 supplement is better than giving an olive oil placebo.

The studies were all small and done over quite a short time period, and not all participants benefitted. So at present it's quite difficult to make recommendations, certainly for the general population. But it appears that omega 3 might one day play a role as an adjunct to treatment for some learning difficulty conditions.

Reducing antisocial behaviour in young people?

Some fascinating studies carried out by scientists from Oxford University have recently looked at the link between nutrition and behaviour in people in young offender institutions. In one study researchers monitored the behaviour of 231 men in Aylesbury's young offender institution for nine months to look at the level of violent incidents. The researchers then randomly divided the group in half. One group received a supplement of vitamins minerals and omega-3 fatty acids. The other group received a placebo pill with no nutrients. Neither the young men nor the staff knew who received which supplement. The researchers monitored behaviour for another nine months during supplementation. To the amazement of all concerned, incidences of violent and antisocial behaviour dropped by 26 per cent, but only in those young men talking the nutritional supplement.

Researchers are currently testing this finding in other similar settings around the UK. If it turns out to be genuine, the finding has major implications for both the possible treatment and prevention of antisocial and violent behaviour in young people.

Chapter 23

Food and Medicine

The science of nutrition emphasises using food to promote health. In other words, a good diet is one that gives you the nutrients you need to keep your body in tiptop condition. However, eating well offers more benefits than simply maintaining normal bodily functions. A good diet may also prevent or minimise the risk of a long list of serious medical conditions including heart disease, type 2 diabetes (an inherited inability to respond to the insulin needed to process carbohydrates), high blood pressure, and cancer.

In this chapter we describe what we know right now about using food to prevent, treat, or cure specific medical conditions. For example:

✔ Eating large amounts of deep green or yellow fruits and veggies coloured with the pigment beta carotene, or red tomatoes and watermelon, which have the red pigment lycopene, may reduce your risk of cancers of the lung, breast, or prostate.

✔ Eating wholegrain cereals high in insoluble dietary fibre (the kind of fibre that doesn't dissolve in your gut) moves food more quickly through your intestinal tract and produces soft, bulky stools that reduces your risk of constipation.

✔ Eating foods such as beans that are high in soluble dietary fibre (fibre that does dissolve in your intestinal tract) seems to reduce the cholesterol circulating in your bloodstream, so preventing it from lodging in and narrowing the walls of your arteries. This reduces your risk of heart disease.

✔ Eating dairy foods (low fat are just as good) and calcium-rich tinned fish, soy products, and green leafy vegetables protects against age-related loss of bone density. It may also help to lower high blood pressure and reduce the incidence of colon cancer.

Not eating certain foods can also be beneficial. Overweight adults who reduce their fat and sugar intake to lose weight may unquestionably prevent a host of medical problems including type 2 diabetes. Similarly, if you're hypertensive, reducing your salt and alcohol intake may help you control your blood pressure.

The joy of food-as-medicine is that it's cheaper and much more pleasant than managing illness with drugs. Given the choice, who wouldn't first opt to try to reduce cholesterol levels with a healthy, low-fat diet, including some oats or beans, rather than with a pill?

Examining Diets with Beneficial Medical Effects

Some foods are so obviously good for your body that no one questions their ability to keep you healthy or make you feel better when you're ill. For example, if you have ever had abdominal surgery, you know all about liquid diets – the clear fluid regimen that your doctor recommended right after the operation to enable you to take some nourishment by mouth without upsetting your gut.

Or if you have type 2 diabetes, you know that your ability to balance the carbohydrate, fat, and protein in your daily diet is important for controlling your blood sugar levels.

Other useful dietary regimens include:

- **The soft diet.** This diet, with lots of minced or puréed foods, is for people who have had head and neck surgery or those who, for any reason – including the result of a stroke – find it difficult to chew or swallow normal-consistency food.

- **The sodium-restricted diet.** A diet low in salt often lowers water retention, which can be useful not only in treating high blood pressure but also congestive heart failure and long-term liver or kidney disease.

- **The cholesterol-lowering diet.** In this plan both total and saturated fat are reduced and you're recommended to eat high-fibre starchy carbohydrates, fruit, vegetables, and unsaturated fats.

- **The low-protein diet.** This diet is prescribed for people who can't metabolise or excrete large amounts of protein. This group includes those with chronic liver or kidney disease or an inherited inability to metabolise amino acids, the building blocks of proteins.

- **The gluten-free diet.** Some people are intolerant to the protein gluten found in wheat, rye, barley, and, to a lesser extent, oats. They may experience bowel problems, weight loss, anaemia, or a severe skin rash as a result. For them a strict gluten-free diet is essential in relieving these symptoms.

Only follow these diets under the supervision of a dietitian.

Using Food to Prevent Disease

Simply adding a missing nutrient to your diet can cure a deficiency disease. For example, scurvy disappears when people eat foods such as citrus fruits high in vitamin C. But what you probably really want to know is whether specific foods or specific diets can prevent illnesses other than deficiency diseases.

This area of nutrition is awash with anecdotes, but anecdotes aren't science. What the nutrition field needs to judge the claims is evidence from scientific studies. Nutritionists examine groups of people on different diets to see how factors such as eating fish, olive oil, fruit, vegetables, and wholegrain cereals (a typical Mediterranean diet), or cutting down on saturated fat, red meat, and salt (more common features of the typical UK diet), can affect the risk of specific diseases.

Sometimes, these studies show a strange effect. For example, a recent study suggested that fruit and vegetables protect against most forms of cancer, although seemingly not breast cancer. Sometimes, studies show no effect at all. And sometimes – we like this category best – they turn up results that nobody expected. For example, in 1996, a study was designed to see whether a diet high in selenium would reduce the risk of skin cancer. After four years, the answer was 'Not so you could notice'. But then researchers noticed that people who ate lots of high-selenium foods had a lower risk of lung, breast, and prostate cancers. Naturally, researchers immediately set up another study, which happily confirmed the unexpected results of the first.

Eating to reduce the risk of cancer

Your daily diet is one of the most important factors in determining the risk of cancer – second only to avoiding tobacco smoke. For example:

- ✔ **Fruits and vegetables.** The active anticancer substances in fruits and vegetables include *antioxidants* (chemicals that prevent molecular fragments called free radicals from hooking up to form cancer-causing compounds) and *phytoestrogens* (hormone-like chemicals in plants that displace natural and synthetic oestrogens in our bodies). For more about these protective substances in plant foods, see Chapter 11.

- ✔ **Foods high in dietary fibre.** Human beings can't digest dietary fibre, but friendly bacteria living in your gut can. Chomping away on the fibre, the bacteria excrete fatty acids that appear to keep cells from turning cancerous. In addition, fibre helps speed food through your body, reducing both the formation and any impact of carcinogenic compounds.

The World Cancer Research Fund (WCRF) *Diet and Health Guidelines for the Prevention of Cancer* are a good start in reducing your risk of cancer. The guidelines are:

✔ Choose a diet rich in a variety of plant-based foods

✔ Eat plenty of fruit and vegetables

✔ Maintain a healthy weight and be physically active

✔ Drink alcohol in moderation if at all

✔ Select foods that are low in fat and salt

✔ Prepare and store foods safely

WCRF Expert Report: Food Nutrition and the Prevention of Cancer: A Global Perspective (1997)

DASHing to healthy blood pressure

In the United Kingdom 16 million people have high blood pressure (also known as hypertension), a major risk factor for heart disease, stroke, and heart or kidney failure. That's one in three of you women readers and two in five of the men.

As you can find out in *High Blood Pressure For Dummies* (published by Wiley), the traditional treatment for hypertension has included drugs (some with unpleasant side effects) combined with specific dietary strategies such as reduced sodium intake, weight loss, alcohol reduction, and regular exercise.

International research known as the 'Dietary Approaches to Stop Hypertension' – DASH, for short – has shown that the overall composition of the diet is very important. The DASH diet is rich in fruit and vegetables, wholegrains, and low-fat dairy products. Poultry, fish, and nuts are the main protein sources. This diet can help prevent an age-related increase in blood pressure and lower an already high level, especially when combined with salt reduction. You can find out more about the DASH diet at www. DashForHealth.com.

Preventing Another Heart Attack

In the UK a total of 1.2 million people have had a heart attack or myocardial infarction at some point in their lives. A heart attack is a complication of coronary heart disease, which is a preventable disease. The death rate from coronary heart disease has been falling since the 1970s, but the UK still has high rates of the disease when compared with other countries. Someone who survives a heart attack is at high risk of suffering another at some point, and a second heart attack is far more likely to kill. Some very simple changes to diet and lifestyle can halp prevent a second heart attack and this has recently led to official guidelines being produced to help reduce the number of second heart attacks.

NUTRITION SPEAK

Vegetarianism: From weird to wonderful

Once upon a time non–meat eaters were regarded as really strange people. Today, vegetarianism is commonplace and, it turns out, pretty good for your health, too. Vegetarianism isn't a single diet. At least three basic variations exist:

✔ People who don't eat red meat but do eat fish and sometimes poultry are known as *demi-vegetarians*. (Strictly speaking people who eat fish or poultry are not true vegetarians.) Around 9 per cent of the UK population fit into this group.

✔ People who don't eat meat, fish, or poultry but do eat other animal products such as eggs and dairy products are called *ovolacto vegetarians* (*ovo* = egg, *lacto* = milk). This group makes up about 5 per cent of the UK population.

✔ People who eat absolutely no foods of animal origin are called *vegans*. These vegetarians, who eat only plant foods, number around 250,000.

The first two regimens are completely safe from a nutritional standpoint because they contain enough different kinds of food to supply every essential nutrient your body needs.

A vegan diet – no animal products at all – is a bit more tricky. It has no natural vitamin B12, a nutrient found only in foods from animals. Vegans therefore need B12 supplements or foods fortified with it such as breakfast cereals, yeast extracts, and meat substitutes.

A vegan diet can also short-change you on calcium and iron. True, many plants have both minerals, but the minerals in the plants are present in forms your body may find harder to absorb (see Chapter 10). And don't forget protein. The proteins in foods from animals are *complete*, meaning that they provide sufficient amounts of all the essential amino acids your body needs to build new tissue, make enzymes, and do all the good things proteins do. Proteins in plant foods, however, are *incomplete* or *limited*, meaning that they provide insufficient amounts of specific amino acids. To build plant-food dishes with complete proteins you need to combine ingredients such as rice and beans, or peanut butter and bread, which *complement* each other, meaning that each provides the amino acids the other needs more of. (For more on proteins, check out Chapter 4.)

In other words, with a little care and planning, you can get all the nutrients you need from a vegetarian diet that may also:

✔ Lower your risk of heart disease (plants are low in saturated fat)

✔ Reduce your risk of some kinds of cancer (those wonderful antioxidant chemicals in plants)

So bring on the vegetable stir-fry with noodles and serve up the rice and beans!

The main dietary issues to help prevent a second heart attack include:

- ✔ Follow a Mediterranean diet.

- ✔ Consume at least 7g of omega 3 fatty acids per week from two to four portions of oily fish, including fresh tuna, sardines, salmon, herrings, trout or pilchards.

- ✔ Consider taking an omega 3 fatty acid supplement for up to four years after a heart attack if you don't eat much fish.

- ✔ Don't take other supplements which contain beta carotene, vitamin C and E or folic acid.

Easing symptoms of the common cold

Let's move on to foods that make you feel better when you have the sniffles – for example sweet foods. Nutritionists know why sweeteners – white sugar, brown sugar, honey, syrup – soothe a sore throat. All sugars are *demulcents*, substances that coat and soothe the irritated mucous membranes. Lemons aren't sweet, and they have less vitamin C than orange juice, but their popularity in the form of warm drinks for colds and sore throat tablets is unmatched. Why? Because a lemon's sharp flavour cuts through to your taste buds and makes the sugary stuff more palatable. In addition, the sour taste makes saliva flow, and that also soothes your throat.

Food as the Fountain of Youth

Can food help you look, feel, or think younger? Certainly some foods provide nutrients that clearly lessen the natural consequences of growing older. For example:

- ✔ Fruits and vegetables rich in antioxidant vitamins may slow the development of cataracts and help prevent age-related macular degeneration (AMD), a major cause of visual impairment in older adults. Eating a variety of different-coloured fruits and vegetables is especially beneficial.

- ✔ Bran cereals give you the fibre that can rev up your intestinal tract. The contractions that move food through your gut slow a bit as you grow older, which is why older people are more likely to be constipated.

Looking after your skin

Some studies suggest that eating well can also offer some protection to your skin. How soon and how much you develop wrinkles depend in large measure on your exposure to the sun (the more sun, the more wrinkling), plus the genes you inherit from your mother and father, but diet plays a role, too. Eating a diet that provides enough calories to maintain a healthy weight won't prevent wrinkles, but it may help you look younger, as people who are underweight can often have saggy skin.

As you get older the *stratum corneum* (the outer layer of your skin) gets thinner and loses its ability to hold moisture. A diet with sufficient amounts of fat won't totally prevent this drying of the skin, but it does give you a measure of protection. Virtually all sensible nutritionists recommend some unsaturated fat or oil every day such as vegetable oils or the omega 3 fats from oily fish.

Some studies suggest that phytochemicals and dietary antioxidants such as vitamin C and selenium protect against sun damage and the formation of age spots (but by no means all dermatologists support this view). We do know that phytochemicals and antioxidants help against many other chronic health problems, so any effect on the skin can only be a bonus!

Keeping your mind young

Recent research shows that older adults who eat a wide range of nutritious foods perform best in memory and thinking tests. Overall good food habits seem to be more important than any one food or nutrient. No one knows for sure right now why and it's not clear whether making lifestyle changes late in life can help or whether you have to eat carefully over a lifetime in order to reap the benefits.

Diet may help to protect against the onset of dementia (where brain cells are damaged and die faster than normal). In the UK about 650,000 people suffer from dementia; a figure that can only increase given our ageing population. Research suggests that high-fat, especially high-saturated-fat, diets may be detrimental, whereas a Mediterranean-type diet containing antioxidants, B vitamins such as folate, fish, and moderate amounts of alcohol may be protective.

French research published in 2002 followed almost 2,000 people over seven years and related their diet to their risk of getting dementia. The researchers concluded that people who ate fish at least once a week had a significantly lower risk of being diagnosed with dementia.

Food and Drug Interactions

Foods nourish your body. Medicines cure or relieve symptoms when something goes wrong. You'd think the two would work together in perfect harmony to protect your body. Sometimes they do. Occasionally, however, foods and drugs compete with each other with more determination than

Olympic athletes. The medicine keeps your body from absorbing or using the nutrients in food or the food (or nutrient) prevents you from getting the benefits of certain medicines.

The medical phrase for this annoying state of affairs is *adverse interaction* or *drug-nutrient interactions*. This section describes several interactions and lays out some simple strategies that make it possible for you to short-circuit them.

How a food and drug interaction happens

When you eat, food moves from your mouth to your stomach to your small intestine, where the nutrients that keep you strong and healthy are absorbed into your bloodstream and distributed throughout your body. Take medicine by mouth, and it follows pretty much the same path from your mouth to your stomach, where it dissolves and passes along to the small intestine for absorption. Nothing unusual about that so far.

Problems can arise when a food or drug brings the process to a screeching halt by behaving in a way that stops your body from using either the drug or the food. Many possibilities exist:

✔ **Some drugs or foods change the natural acidity of your digestive tract so that you absorb nutrients less efficiently.** For example, your body absorbs iron best when the inside of your stomach is acidic. Taking antacids reduces stomach acidity – and iron absorption.

✔ **Some drugs or foods change the rate at which food moves through your digestive tract, which means that you absorb more (or less) of a particular nutrient or drug.** For example, eating prunes (a laxative food) or taking a laxative drug speeds things up so that foods (and drugs) move more quickly through your body and you have less time to absorb medicine or nutrients.

✔ **Some drugs and nutrients bond to form insoluble compounds that your body can't break apart.** As a result, you get less of the drug and less of the nutrient. The best-known example: Calcium (in dairy foods) bonds to the antibiotic tetracycline so that both zip right out of your body without giving you time to absorb them.

✔ **Some drugs and nutrients have similar chemical structures, and taking them at the same time fools your body into absorbing or using the nutrient rather than the drug.** One good example is warfarin (a drug that keeps blood from clotting) and vitamin K (a nutrient that makes blood clot). Eating lots of vitamin K-rich leafy green vegetables counteracts the intended effect of taking warfarin.

✔ **Some foods contain chemicals that either fight or intensify the natural side effects of certain drugs.** For example, the caffeine in coffee, tea, and cola drinks reduces the sedative effects of antihistamines and some antidepressant drugs, while increasing nervousness, insomnia, and shakiness common with cold medications containing caffeine or a *decongestant* (an ingredient that temporarily clears a stuffy nose).

Food fights: Drugs versus nutrients versus drugs

In the mid-1990s, researchers tracking the effects of alcohol beverages on the blood pressure drug felodipene stumbled across the *grapefruit effect*, a dramatic reduction in your ability to metabolise and eliminate certain drugs if you take them with grapefruit juice. The result may be an equally dramatic rise in the amount of medication in your body, leading to unpleasant side effects. Ooops. Since then, the list of drugs that interact with grapefruit juice has expanded beyond felodipine to include a second blood pressure medicine, nifedipine, plus – among others – the cholesterol-lowering drugs lovastatin, pravastatin, and simvastatin; the antihistamine loratadine; the immunosuppressant drug cyclosporine; and saquinivar, a protease inhibitor used to treat HIV.

The offending ingredient in grapefruit juice remains a mystery, but one leading candidate is *bergamottin*, a naturally occurring compound that inactivates a digestive enzyme needed to convert many drugs to water-soluble substances you can flush out of your body. Without the enzyme, you can't get rid of the drug. Double ooops. By the way, if you feel particularly wired after drinking grapefruit juice with your morning coffee or tea, maybe it's because the juice also interacts with caffeine. Who could have guessed?

Other food and drug fights are in the following list, and in Table 21-1.

✔ Water pills, more properly known as *diuretics*, make you urinate more often and more copiously, thus increasing your elimination of the mineral potassium. To make up what you lose, experts suggest adding potatoes, bananas, oranges, spinach, corn, and tomatoes to your diet. Consuming less sodium (salt) while you're using water pills makes the water pills more effective and decreases your loss of potassium.

✔ Oral contraceptives seem to reduce the ability to absorb B vitamins, including folate. Taking lots of aspirin or other NSAIDS (nonsteroidal anti-inflammatories) such as ibuprofen can trigger a painless, slow but steady loss of small amounts of blood from the lining of your stomach that may lead to iron-deficiency anaemia.

✔ Persistent use of antacids made with aluminum compounds may lead to loss of the bone-building mineral phosphorus, which binds to aluminum and rides right out of the body. Laxatives increase the loss of minerals (calcium and others) in faeces.

✔ The antiulcer drugs cimetidine and ranitidine can make you positively giddy. These drugs reduce stomach acidity, which means that the body absorbs alcohol more efficiently. According to experts, taking ulcer medication with alcohol leads to twice the wallop, like drinking one beer and feeling the effects of two.

Read the label on your medication carefully and check with your doctor or pharmacist for any potential food and drug interactions whenever you take medication. Remember that your medication may also interfere with any nutritional supplements you take. The vitamins and minerals in nutritional supplements are simply food reduced to its basic nutrients, so frequent interactions between drugs and supplements shouldn't be too surprising.

Table 21-1	Medication Most Likely to Have Nutritional Implications
Medication	*Common side effects on nutritional intake*
Amphetamines (nervous system stimulants)	Increases appetite
Analgesics (pain killers)	Anti-inflammatory analgesics NSAIDs* can cause gastric irritation
Antacids (counter stomach acidity)	Binds with minerals such as iron, zinc, and phosphorus, reducing their absorption
Antibiotics (kill bacteria)	Can alter gut flora and cause diarrhoea. Tetracycline binds with calcium, reducing the absorption of both the drug and the nutrient
Anticoagulants (reduce blood clotting)	Warfarin action is affected by vitamin K
Anticonvulsants	Interferes with folate and vitamin D metabolism
Antidepressants	Can cause weight loss, MAOI** type necessitates avoidance of amino acid tyromine
Antihyperlipidaemics (help control fats in blood)	Clofibrate and cholestyramine can result in malabsorption of minerals and fat-soluble vitamins
Antipsychotics	Some types significantly increase appetite
Corticosteroids (anti inflammatory drugs)	Weight gain and glucose intolerance

Medication	Common side effects on nutritional intake
Cytotoxic (anticancer drugs)	Can lead to weight loss, taste changes, nausea, vomiting, damage to the intestinal lining, poor folate metabolism, reduced thiamin status
Diuretics (help remove excess fluid from body)	Thiazide and loop diuretics can cause excessive urination and losses of potassium, calcium, zinc, and water-soluble vitamins Dehydration
Hypoglycaemics (help control blood sugars)	Drug action has to be balanced with carbohydrate intake. Alcohol can make drug more potent and can cause facial flushing
Laxatives (treat constipation)	Can reduce nutrient absorption
Mood stabilisers	Lithium effectiveness influenced by dietary sodium
Oral contraceptives	Affects glucose and lipid metabolism and can cause weight gain

*NSAID: non steroidal anti-inflammatory drug; **MAOI: monoamine oxidase inhibitor
Source: Adapted from Thomas, B. The Manual of Dietetic Practice 3rd Ed. 2004 Blackwell Publishing Ltd. Oxford

The Last Word on Food versus Medicine

Sometimes, people with a life-threatening illness are worried by the side effects of drugs or the lack of certainty in standard medical treatment. In desperation, they may reject conventional medicine and accepted drug therapy and turn instead to alternative dietary treatments. Alas, doing this may be hazardous to their already-compromised health. When someone is ill, a nutritious diet not only promotes physical recovery, helping to fight infection and heal wounds, it can also improve morale and mood. So food may enhance the effects of treatment but no one has found that it serves as an adequate, effective substitute for (among others) the following drugs:

✔ Antibiotics and other medication used to fight infections including HIV

✔ Vaccines or immunisations used to prevent infectious diseases

✔ Anticancer drugs used in chemotherapy

If a health professional suggests altering your diet to make your treatment more effective, that makes sense. But if someone suggests abandoning your conventional medical treatment in favour of an alternative dietary therapy, discuss the pros and cons with your doctor first. Some unconventional diets can be at best unpleasant and difficult to follow, and at worst detrimental to your health – as yet, no truly magical foods exist.

Chapter 24

Dietary Supplements

· ·

· ·

*T*he multivitamin pill you may pop each morning is a dietary supplement. So are calcium tablets, garlic capsules and cod liver oil. Echinacea, a herb reputed to boost the immune system, is one, and so too is glucosamine sulphate, which many people take for their joints. Other commonly used supplements include evening primrose oil, omega-3 fatty acids, antioxidants and iron tablets.

Dietary supplements can be single-ingredient products, such as vitamin E capsules, or they may be combination products, such as multivitamin pills or the enriched amino acids and protein powders that some athletes favour.

An estimated 10 million adults in the United Kingdom take dietary supplements on a regular basis. A recent survey showed that around one in five people report using them at some time or another, with around half of British women over 50 taking at least one dietary supplement each day. The average spend on dietary supplements is approximately £100 per person a year. Herbal remedies are also big business, with one in five adults using them. In 2009 the UK retail market for dietary supplements was worth £364 million – of which vitamin and mineral supplements account for £130 million. Multivitamins make up 25 per cent of all supplement sales, vitamin C on its own 14 per cent and mineral supplements 8 per cent.

You can stir up a healthy debate in any group of nutrition experts simply by asking whether all these supplements are (a) necessary, (b) economical or (c) safe. But when the argument's over, you still may not have the definitive answer, so in this brief chapter we aim to provide you with the information you need to make your own best informed choices.

Looking at Why People Use Dietary Supplements

In a country where food is plentiful and relatively cheap, you may wonder why so many people opt to pop pills instead of just plain food. Health-conscious individuals, most often women and older adults, frequently take multivitamins as a preventive measure. People often take single nutrients in the hope of preventing or treating a specific illness, such as vitamin C for colds, vitamin B6 for premenstrual syndrome (PMS) or herbal valerian for insomnia.

A recent survey by the Medical Research Council found that most of the people who take food supplements are those on a higher income with access to a more nutritious diet or those with a good level of knowledge and motivation to follow a healthy lifestyle including not smoking and taking more exercise. In other words, people who take food supplements are often those least likely to need them!

In general, nutrition experts, including the Food Standards Agency and the British Dietetic Association, recommend that you invest your time and money in eating meals and snacks that supply the nutrients you need in a balanced, tasty diet. Nonetheless, every expert worth his or her vitamin C admits that in certain circumstances, supplements can be useful.

Vitamin-deficiency diseases such as scurvy and beriberi are rare in Western countries. But *suboptimal vitamin levels* – sciencespeak for slightly less than you need – can be problematic. If 'slightly less than you need' sounds slightly less than important, consider this:

- Suboptimal intakes of folate and vitamins B6 and B12 raises the risk of birth defects such as spina bifida in an unborn foetus, and are believed to be linked to heart disease and possibly colon cancer.

- Suboptimal vitamin D intake means poor bone health (a higher risk of rickets, osteomalacia or osteoporosis) and possibly a higher risk of breast, ovarian, prostate and colon cancers.

- Suboptimal levels of antioxidant vitamins A, E and C together with selenium are linked to heart disease, some forms of cancer and possibly other diseases linked to increasing age such as dementia, cataracts and macular degeneration of the eyes.

Supplementation with individual antioxidants either on their own or in combination has had disappointing results, suggesting that supplements aren't a true reflection of the protective mix you get from eating five portions of fruit and vegetables a day. More worryingly, recent work in Denmark looking at the results of a number of clinical trials to assess the overall effects of supplements has shown that treatment with beta-carotene, vitamin A and vitamin E may actually be bad for your health. The results show that in 47 trials involving 180,938 people, antioxidant supplements actually increased death rates by 5 per cent. When they looked at individual vitamins, beta-carotene increased mortality by 7 per cent, E by 4 per cent and A by 16 per cent.

In the UK guidelines on heart health now reflect these findings and specifically advise that heart patients shouldn't take supplements containing beta-carotene, and shouldn't be advised to take antioxidant supplements (vitamin E and/or C) or folic acid to reduce cardiovascular risk. In addition, people who are at increased risk of lung cancer, such as smokers or those who have been exposed to asbestos, shouldn't take any beta-carotene supplements. Head to Chapter 9 for all the news on vitamins.

When food isn't enough: Using supplements as insurance

Illness, age, dietary restrictions and some gender-related conditions may put you in a spot where you can't get all the nutrients you need from food alone, either because you're eating a less balanced diet or because your needs are greater than normal. In such cases short- or long-term intake of supplements may prove useful.

Digestive illnesses, unfriendly drugs, injury and chronic illness

Certain metabolic disorders and diseases of the digestive tract and major organs (such as coeliac disease or liver or kidney disease) interfere with the normal digestion of food and the absorption or metabolism of nutrients. Some medicines may also interfere with normal digestion, meaning that you need supplements to make up the difference. People who suffer from certain chronic diseases, who've suffered a major injury (such as a serious burn) or who've just been through surgery may need more nutrients than they can get from food. In these cases, a doctor may prescribe supplements to provide the nutrients.

Vegetarianism

Vitamin B12 is found only in food from animals, such as meat, milk and eggs. Without these foods, *vegans* – people who don't eat any foods of animal origin – will almost certainly have to get their vitamin B12 from supplements or from specially fortified foods such as breakfast cereals, meat substitutes or yeast extracts. A calcium supplement may be useful for vegetarians who don't eat dairy products or another good source of calcium (such as fortified soya milk, sesame seeds or tofu). Vegetarians who eat some animal foods may get their B12 and calcium but may still be short of zinc and iron. You find these minerals in some plants but in a form your body doesn't absorb as easily as the kind found in animal foods. Supplements can help plug the gap.

Children

Growing children have high nutrient needs. Vitamin drops containing vitamins A, C and D are available for children from 6 months old to 5 years. Healthy Start vitamin drops are free to parents on low incomes.

Athletes

Food supplements won't enhance your performance if your diet is already adequate, and supplements are certainly no substitute for good training and a healthy lifestyle. However, for athletes who need to restrict their diet to maintain a low body weight (including some gymnasts, dancers and distance runners) multivitamins together with iron and calcium can be useful. Antioxidants may help counteract some of the muscle damage arising from strenuous exercise, but you can get sufficient antioxidants from including plenty of fruit and vegetables in your diet.

An enormous range of food supplements is marketed to athletes as performance enhancers (so-called *ergogenic acids*). These include protein powders and bicarbonate, as well as herbal remedies such as guarana and ginseng. Unfortunately, credible conclusive scientific evidence of the effects of these products is lacking – along with an understanding of the long-term health risks and short-term side effects.

Creatine is an ingredient in some performance-enhancing products. It's an amino acid found in meat and fish and also made in the body. It plays a role in the muscles, generating energy for short bursts of activity in multiple sprint type sports such as football. But creatine has no benefit in other endurance type sports and can lead to a short-term weight gain, which can impair performance for some athletes.

Older adults

As you grow older, your appetite may decline and your sense of taste and smell may falter. If food no longer tastes as good as it once did; if you have to eat alone all the time and don't enjoy cooking for one; or if dentures make chewing difficult, you may not be taking in the variety of foods that you need. To get the nutrients you require dietary supplements can be useful, especially water soluble vitamins such as folate and vitamin C along with iron, zinc and calcium. Look for a multivitamin and mineral that contains your daily requirement in one tablet or ask your GP for advice.

Recent studies show that a third to a half of elderly people suffer from a lack of vitamin D during the winter months. If you're elderly and housebound (or have little exposure to the sun) you may benefit from a daily supplement of 10 micrograms of vitamin D or a larger dose that you can receive in an injection that lasts several months. Once again, ask your GP for advice.

Women

At various stages of their reproductive lives, women benefit from supplements as insurance:

- **Before menopause:** Women with heavy periods can lose a lot of iron each month. They rarely get sufficient amounts of iron to compensate from a typical diet providing fewer than 2,000 calories a day. For them, and for women who are dieting to lose weight, iron supplements may be the only practical answer.

  Some women take vitamin B6 supplements to help alleviate premenstrual syndrome (PMS), although a clear benefit is as yet unproven. But take it easy: doses of vitamin B6 above 50 milligrams a day are linked to symptoms of (thankfully reversible) nerve damage manifested as numbness and weakness in the hands and feet.

- **Pregnancy:** Women planning to get pregnant and up to week 12 of gestation are advised to take 400 micrograms of folic acid per day to reduce the risk of neural tube defects (such as spina bifida) in the developing foetus by up to 70 per cent. Pregnant women on low incomes are entitled to Healthy Start vitamin supplements containing folic acid, vitamin C and D. Some pregnant and breastfeeding mothers are also advised to take 10 micrograms of vitamin D daily, and pregnant and breastfeeding women may also be advised by their prenatal healthcare team to take other supplements such as iron, to provide the nutrients they need to build new maternal and foetal tissue.

  Never self-prescribe supplements while you're pregnant. Consult your doctor beforehand and take only pregnancy-formulated supplements unless specifically advised otherwise by your doctor. Large amounts of some nutrients may actually be hazardous for your baby. For example, taking megadoses of vitamin A while you're pregnant can increase the risk of birth defects.

- **Bone health:** Calcium and vitamin D are important for those at risk of osteoporosis. Most adult women get the calcium they need (700 milligrams per day) from two to three servings of dairy produce (boosted by the small amounts coming from tinned fish, nuts and seeds, soya, pulses, dried fruit and hard water). Achieving this nutritional balancing act every single day may be unrealistic, particularly if you don't eat dairy products. An alternative is calcium supplements. Adequate calcium is just as important for healthy bones in men.

Using supplements instead of medicine

Here's the lowdown on the latest thinking around taking supplements to relieve medical conditions:

✓ **Black cohosh:** A recent study in America – the Herbal Alternatives for Menopause Study – found that black cohosh, another of the major supplements purchased in the UK as a natural alternative to HRT, was no better than placebos for relieving the frequency or intensity of hot flushes in women during the menopause.

✓ **Ginkgo biloba:** Extracted from leaves of the maidenhair tree, many people take this supplement for a wide range of conditions, but evidence supports a minor but positive role in only two: reducing pain during walking resulting from a narrowing of the arteries in the leg (intermittent claudication) and a moderate relief of some of the symptoms of dementia.

✓ **Echinacea:** Did you know that no medical cure for the common cold exists? Through the years, several supplements, such as zinc lozenges, have been touted as cold remedies. Most have fallen by the wayside, but many people use a new contender, the herb echinacea (pronounced e-kin-ay-sha), on which the British public currently spend £1 million per month to try and treat the symptoms of or prevent colds. The most recent studies have shown that insufficient evidence exists to recommend echinacea for this purpose, although one study did show some very modest relief in the severity of symptoms.

✓ **Glucosamine and chondroitin sulphate:** Many people take glucosamine and/or chondroitin sulphate supplements for joint problems such as osteoarthritis or a sports injury to halt the breakdown of cartilage and reduce inflammation. Not all experts are in agreement, but some convincing research shows that both glucosamine and chondroitin, taken separately or together, may help alleviate stiffness, pain and inflammation as well as improve range of movement in some people with joint problems. But you may need to take the supplements for at least a month to see a benefit.

✓ **Omega-3 fish oil:** Fish body oils rich in omega 3 are used for joint tenderness and morning stiffness in rheumatoid arthritis, heart disease and even learning problems in children. Some interesting ongoing studies have shown a benefit for these conditions, although others say otherwise. Use fish body oils rather than fish liver oils to avoid overdosing on vitamins A and D.

✓ **St John's wort:** This herb is widely used as an antidepressant in cases of mild to moderate depression. Some studies have shown it to be more effective than a placebo, with potentially fewer side effects than conventional antidepressants. However, you can't take St John's wort together with a wide range of prescription drugs, including oral contraceptives. Despite a large number of trials, whether the herb really is a safe and effective treatment and which doses and extracts are most effective remains unclear.

Supplement Safety: An Iffy Proposition

In the United Kingdom the range of products on sale is greater and the levels of ingredients found in them higher than in many other countries where upper limits on particular ingredients are strictly enforced. Some supplements sold under food law in the United Kingdom can only be sold as medicines in other countries. This could mean that these supplements have not yet been proven as effective or safe and may need to be viewed with caution, as their quality and efficacy are not necessarily known.

In May 2003 a report from the United Kingdom's Expert Group on Vitamins and Minerals (EVM) set safe upper levels for a variety of micronutrients in supplements. The majority of supplements do comply with these levels. However, a small number with high contents of iron, calcium, magnesium, nickel, zinc, vitamin A, vitamin B6, beta carotene and nicotinic acid are a cause for concern. Risks may be especially high if consumers take multiple supplements and inadvertently double up on nutrients. The EU is currently discussing how to achieve the difficult balance between consumers needs and wishes and possible health risks from excessive intakes of these micronutrients.For more on the safe and recommended levels for vitamins and minerals, see Chapters 9 and 10.

Herbal remedies, some with relatively little supporting scientific evidence for their claims, are also sold in UK without a medicines licence. Some herbal remedies contain strong chemicals that can have side effects or interact dangerously with prescribed drugs. As a general rule, the safety data isn't sufficient torecommend that you use these products during pregnancy, breastfeeding or for children, and do consult your GP before use.

Sweet trouble

Nobody wants to force down a foul-tasting supplement, but pills that look or taste like sweets may be hazardous to a child's health. Some nutrients are troublesome – or even deadly – in high doses (see Chapters 9 and 10), especially for kids. You're experienced enough to know not to triple your dose just because the supplement tastes like strawberries, but you can't count on a child to be that sophisticated. If you have youngsters in your house, protect them by buying neutral-tasting supplements and keeping them in a safe cabinet, preferably high and locked tight to resist tiny prying fingers.

Just because herbal supplements are 'natural' doesn't make 'em safe! Cases of fatalities have been linked to some herbal remedies and kava kava and ephedra are now banned from sale in the United Kingdom. Some, such as black cohosh (used to relieve PMS), chaparral, kombuchu tea, comfrey and pennyroyal, can have potentially worrying side effects in some people. Other widely used herbal supplements, including aloe vera, echinacea, ginkgo biloba and ginseng, are generally considered safe but can occasionally have side effects (see Table 22-1).

A very welcome European Union directive on traditional and herbal medicines came into force in 2004 to help protect consumers. By 2011 the directive will ensure the purity and safety of herbal remedies, the safety and maximum permitted doses, what conditions they may be effective in treating and what precautions you should consider before taking them.

Table 22-1	Common Herbal Remedies	
Herb	*Purpose*	*Known Problems and Reported Side Effects*
Aloe vera	For digestive disorders	Occasional mild itching or a rash; laxative effect
Echinacea	Protects against infections and boosts the immune system	Long-term safety unclear especially in immuno-suppressed individuals (where your immune system works less effectively); rare allergic reactions reported
Ephedra or ma huang	Stimulant used for weight loss and to boost athletic performance	Banned in UK and US following links to fatalities from heart disease and stroke; side effects of anxiety and insomnia as well as raised blood pressure
Evening primrose oil	For breast pain and PMS	Rare but may include headache, nausea and stomach ache; avoid if you suffer from epilepsy
Garlic	To reduce infections, blood pressure, cholesterol levels and cancers of the digestive tract	Gastric irritation, indigestion and nausea; potentially dangerous if taken with blood-thinning medication or before surgery

Herb	Purpose	Known Problems and Reported Side Effects
Ginkgo biloba	To improve symptoms associated with dementia, memory loss, circulatory problems, sexual problems and tinnitus	Few reported side effects but can have mild gastrointestinal effects; allergic skin reactions reported; potentially dangerous if taken with blood-thinning medication, some psychiatric drugs and blood sugar-controlling medication, or before surgery
Ginseng	As a tonic to improve wellbeing, performance, memory and stamina	Insomnia, diarrhoea, headache and mild oestrogen related hormonal effects in post-menopausal women; potentially dangerous if taken with blood-thinning medication
Guarana	To boost alertness and reduce fatigue; as a weight loss aid	Similar effects to large doses of caffeine – nervousness, anxiety, insomnia and palpitations; possible rise in blood pressure
Kava kava	Natural tranquilliser	Liver damage (banned following a link to several fatal cases)
Royal jelly	As a tonic to help general health	Allergic reactions, especially in those with a history of asthma or allergy
Saw palmetto	For prostate and urinary tract problems	Rare but may include stomach upsets
St John's wort	For mild to moderate depression and anxiety	Interactions with other medication including other antidepressants, blood-thinning medication, immuno-suppressants, some asthma, migraine and epilepsy medications and oral contraceptives; hypersensitivity to light
Valerian	For insomnia, anxiety and irritable bowel	Excessive tiredness; withdrawal symptoms

Source: Adapted from The Consumers' Association (2001) and Dietary Supplements & Functional Foods by G Webb (2006)

Choosing the Most Effective Supplements

Okay, you've read about the virtues and drawbacks of supplements. You've decided which supplements you think may do you some good. Now it's crunch time, and all you really want to know is how to choose the safest, most effective products. It's supplementary, my dear Watson! The guidelines in this section should help.

Pick a well-known brand

Even though food supplement manufacturers aren't required to prove effectiveness, a respected name on the label should offer some assurance of quality and purity of the product. This doesn't always mean that the most expensive supplement is the best – many supermarkets and pharmacy chains supply excellent own-brand food supplements at a reasonable price.

Check the ingredient list

Check the supplement label (for more about the nutrition labels, see Chapter 14). Supplement labels should list all ingredients. The label for vitamin and mineral products must give you the quantity per nutrient per serving plus the percentage of the recommended daily amount (RDA) Figure 22-1 shows some of the information to look out for on a good-quality supplement label.

Look for the expiry date

Over time, all dietary supplements become less potent. Always choose a product with a long shelf life. Skip the ones that will expire before you can use up all the pills, such as the 100-pill bottle with an expiry date 30 days from now.

Check the storage requirements

Even when you buy a product with the correct expiry date, it may be less effective if you don't keep it in the right place. Some supplements must be refrigerated; the rest you need to store, like any food product, in a cool, dry place.

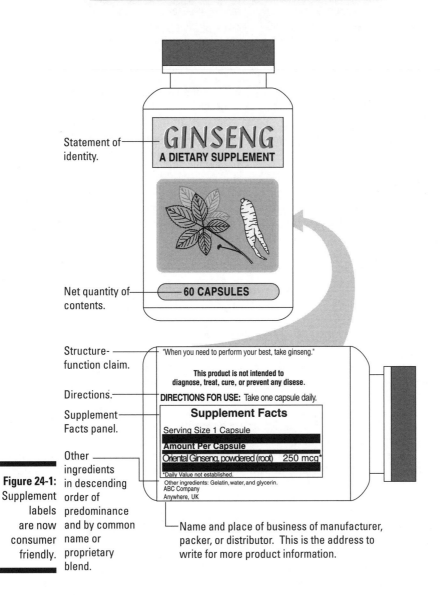

Nutrition Labeling for Dietary Supplements

Statement of identity.

GINSENG
A DIETARY SUPPLEMENT

Net quantity of contents.

60 CAPSULES

Structure-function claim.

"When you need to perform your best, take ginseng."

This product is not intended to diagnose, treat, cure, or prevent any disese.

Directions.

DIRECTIONS FOR USE: Take one capsule daily.

Supplement Facts panel.

Supplement Facts

Serving Size 1 Capsule

Amount Per Capsule

Oriental Ginseng, powdered (root)   250 mcg*

*Daily Value not established.

Other ingredients in descending order of predominance and by common name or proprietary blend.

Other ingredients: Gelatin, water, and glycerin.
ABC Company
Anywhere, UK

Name and place of business of manufacturer, packer, or distributor. This is the address to write for more product information.

Figure 24-1: Supplement labels are now consumer friendly.

Energy drinks under the microscope

In recent years a whole range of soft drinks marketed as 'energy drinks' or 'stimulant shots' have exploded onto the UK market. They claim to boost your energy levels, alertness concentration and even academic or sporting performance. Energy drinks usually contain high levels of caffeine and may also contain other stimulants, such as guarana, ginseng, taurine or glucuronolactone, and sometimes vitamins and minerals or other herbal substances alongside calories or energy as sugar.

For some, the caffeine levels in a 250-milliletre can are usually about the same as you find in a mug of strong coffee. However, some of the more recent arrivals on the UK market from the US are around four times this strong, giving the equivalent of two strong mugs of coffee in a 60-milliletre shot. At present the label must inform the consumer that the drink is high in caffeine and give the amount per 100 grams. Many do also state that the drinks are unsuitable for children and pregnant women, but no regulations limit the caffeine content.

This concerns many health experts because large doses of caffeine can cause a variety of unpleasant symptoms including irritability, insomnia, anxiety, tremors, palpitations and nausea, especially in those who aren't used to regular high caffeine intakes. A recent study in the Journal of Drug and Alcohol Dependence highlighted the cases of nine teenagers admitted to hospital in the US with symptoms of caffeine intoxication following consumption of energy drinks, and called for the drinks to carry warnings highlighting these risks.

Choose a sensible dose

Unless your doctor prescribes a dietary supplement as medicine, you don't need products marked 'therapeutic', 'extra-strength' or any variation thereof. Pick one that gives you no more than the recommended daily amount (RDA) for any ingredient. Luckily, you don't have to memorize RNIs to know which products these are. Just look for the percentage of the daily value the supplement provides. For example, the *DV* (daily value) for vitamin C is currently 40 milligrams for an adult. A product containing 40 milligrams of vitamin C provides 100 per cent DV and supplies all the vitamin C an adult needs in a day.

Beware of the health claims on supplements

When the label promises something that's too good to be true, it probably is! At present, United Kingdom food law doesn't require food supplements to demonstrate their efficacy before marketing. So labels can make claims of health benefit – 'calcium can help build strong bones' or 'garlic can help healthy circulation' – as long as generally accepted evidence from nutrition research exists of a health benefit in humans. However, supplement marketers can't claim that their products cure or prevent a disease nor a symptom of a disease (such medicinal claims are only allowed on licensed medicines).

Confused? So are we. The good news is this legislation is currently under review, which should make it easier for the consumer to sort the fact from the fiction.

Good Reasons for Getting Nutrients from Food Rather than Supplements

Despite this chapter's focus on the sensible use of supplements, we feel obliged to play devil's advocate and report to you the arguments in favour of healthy people getting all or most of their nutrients from food rather than supplements.

Cost

If you're willing to plan and prepare nutritious meals, you can almost always get your nutrients less expensively from a varied and balanced diet. This means plenty of fresh fruit, vegetables, wholegrain cereals, dairy products, meat, fish and poultry. Besides, food usually tastes better than supplements.

Unexpected bonuses

Food is a package deal containing vitamins, minerals, protein, fat, carbohydrates and fibre, plus a cornucopia of plant chemicals (phytochemicals) that may be vital to your continuing good health. Think of lycopene, the red pigment in tomatoes that was recently found to reduce the risk of prostate cancer. Think of genistein and daidzein, the oestrogen-like substances in soya beans that appear to reduce your risk of heart disease. Who knows what else is hiding in your apples, peaches, pears and plums? Do you want to miss out on these goodies? Of course not. For more about the benefits of phytochemicals, see Chapter 11.

Safety

Several common nutrients may be toxic if you take them in *megadose servings* (amounts several times larger than the recommended daily allowances). These effects are more likely to occur with supplements. Pills slip down easily, but regardless of how hungry you are, you probably won't eat enough food to reach toxic levels of nutrients, with the exception of Vitamin A. (To read more about the hazards of megadoses, see Chapters 9 and 10.) If you are going to take a supplement, look on the label for a general multi-preparation with no more than 100–150 per cent of the recommended daily allowance unless advised otherwise by a health professional.

Part VI
The Part of Tens

'The koi carp have gone again – that's the
<u>very</u> last time you bring your Dieting &
Nutrition Club round for afternoon tea!'

In this part . . .

*1*f you've ever read a *For Dummies* book, you know what to expect here — nifty lists each containing essential information about a related topic.

In this book, that means ten great nutrition-related Web sites, the truth behind ten trendy diets, and the low down on ten superfoods. Who could ask for anything more?

Chapter 25

Ten Nutrition Websites
You Can Trust

. . . .

In This Chapter

▶ Surfing the web for balanced, up-to-date information on nutrition

▶ Finding answers to your questions about diet and health

▶ Staying abreast of the latest nutritional hot topics

. . . .

*H*ow did we ever manage without the Internet? Every time we open our e-mail or download an article from a journal that would have taken us hours to locate in the prehistoric days *BTW* (before the web), we feel a genuine need to say, 'Thank you, technology!' The ten websites we list in this chapter give you reliable, accurate balanced information: nutritional guidelines, news stories, practical tips, interactive sites, directories and more right at the end of your fingertips – just go where the fancy takes you.

British Dietetic Association

www.bda.uk.com

This site features nutrition recommendations, guidelines, press releases, research and policies from the UK's membership association for registered dietitians. (For a quick rundown on who's who in nutrition science, see Chapter 1.)

The BDA home page features links to categories that are clearly meant to appeal to association members. However, as a consumer you can also find a wealth of information on how to train as a dietitian, and the wonderful 'Food facts', a range of well-written and authoritative information sheets on key areas of nutrition. The strength of these fact sheets lies in the translation of evidence-based science into practical advice for the consumer. What's more, you can download and print them off for your own use or even see links to the original scientific studies on which they were based if you want more information. We counted over 60 fact sheets last time we looked – including hot topics such as the glycaemic index, fad diets and detoxing, probiotics, whole grains and irritable bowel syndrome.

Look no further if you want a reliable, unbiased source of information on the more nuts-and-bolts areas of nutrition such as vegetarianism, food allergies, food supplements and weaning.

From the main website you see a link to the BDA's consumer websites dedicated to weight loss for adults (www.bdaweightwise.com) or for teenagers (www.teenweightwise.com). These sites contain a wealth of useful, evidence-based, practical suggestions for all stages of losing weight and keeping it off.

The main BDA site has information on how to access an NHS dietitian. You can also check from the home page whether a particular dietitian is registered to practise with the Health Professions Council or find a freelance dietitian by clicking on the link to:www.freelancedietitian.org. This site provides details of dietitians who work in a broad range of freelance activities, including many who provide private consultations.

British Nutrition Foundation

www.nutrition.org.uk

The British Nutrition Foundation is a charity offering sound nutritional information and advice to scientists and the public alike. The website contains all the key information you need on nutrition through the life stages, in health and illness. We like the 'Food commodities' section that lets you browse through a range of your favourite foods. You can also discover a great nutrition education programme intended for pupils, teachers and parents with masses of useful resources you can download. The section on publications enables you to order unbiased, well-produced leaflets, posters, software, videos and books on nutrition.

Diabetes UK

www.diabetes.org.uk

Diabetes UK is the nation's largest organisation working for people with diabetes, funding research, campaigning and helping people live with the condition. You could spend a whole day just cruising around their site for sound information on how to manage diabetes. The dietary advice on healthy eating (see 'Food and recipes' in the 'Guide to diabetes' section) applies to most people, whether they have diabetes or not. The site has great practical tips for cooking, meal planning, choosing snacks, eating out and weight management, as well as special diabetes-related topics such as low blood sugar, diabetes foods, sweeteners and what to eat if you're unwell or doing physical activity. You can even do a virtual shopping tour online – amazing! And information on diet is available to download in an incredible 21 different languages!

Food and Drink Federation

www.fdf.org.uk

The Food and Drink Federation is the voice of the UK food and drink manufacturing industry. The FDF website is a mine of information about the industry. You can scan the home page for interesting food-related news stories and hot topics such as trans fats and food wastage. It also helps you follow the interesting progress some manufacturers are making towards developing healthier products with lower salt, fat and sugar levels.

Some coverage of nutritional topics is limited, but other coverage is comprehensive – for instance, the linked section on Guideline Daily Amounts (GDAs) and food labels gives you all the information you could ever want to know and more. Go to:

http://www.whatsinsideguide.com

Food Standards Agency

www.eatwell.gov.uk

With so many sources of information on the Internet, you may find yourself stranded at sea in an ocean of information – maybe even some misinformation. The Food Standards Agency's consumer-faced website, Eat Well, Be Well, enables you to steer through the waves to a safe harbour.

This website covers aspects of healthy eating, food safety, technology, hygiene and nutrition. From the home page you can access dietary recommendations for different age groups and medical conditions, as well as key consumer issues such as food labelling. The site presents data in a fun but informative way. You can also find bang up-to-date sections on news stories and food scares from GM foods to additives. If you fancy a bit of fun, this site has interactive games and quizzes, including your very own Body Mass Index and calorie expenditure calculators. We also love the 'Ask the expert' section, covering everything you ever wanted to know about nutrition but didn't know who to ask!

Grub4Life

www.grub4life.org.uk

This website was founded by one of us (ND) so we make no apologies for including it in our top ten because we hope you'll love it as much as we do. The site has everything you could possibly want to know about feeding children in the early years. You can read all the latest news or watch online videos of delicious, healthy recipes road-tested by thousands of youngsters. You can even join a chat forum if you become a buddy of the site.

Heart UK / British Heart Foundation

www.heartuk.org.uk / www.bhf.org.uk

Both Heart UK and the British Heart Foundation offer advice and support to people who want to reduce their risk of heart disease. The websites offer a wealth of good information on heart disease, including the role of lifestyle factors such as diet.

You can download free leaflets on a heart-healthy diet or scroll through specific information sheets on a wide range of heart health topics, including how to reduce saturated or salt fat in cooking, how to eat out healthily, what to put in your own or your children's lunch boxes, the benefits of oily fish or alcohol and how vitamins affect heart disease. The sites have wonderful heart-healthy recipes and links to online videos proving useful information.

NHS Choices – Live Well

www.nhs.uk/livewell

The Live Well section of the NHS Choices website has literally hundreds of topics to choose from including some really good sections on healthy eating, weight loss and alcohol. We also like the more unusual coverage of diet related topics such as digestive disorders, vegetarianism, eating disorders and food allergies. The site has a great dietary self-assessment tool and a link into a five-a-day section that contains all you could want to know and more about fruit and vegetables.

United States Department of Agriculture Nutrient Database

www.nal.usda.gov/fnic/foodcomp/search

The USDA Nutrient Database is the ultimate food list, with data for more than 5,000 commonly eaten foods. Each entry is a snapshot of the amount of all the main nutrients in a specific food serving (a raw apple, for example). Ever wondered what's in that Danish pastry? With this site you're just a couple of clicks away from finding out.

Type the name of the food you're looking for in the text box on the home page – Danish pastry, for example – then click Submit. A new screen pops up listing various serving sizes. Click on the box in front of the serving sizes you want to look at, click Submit again, and – bingo! – calories, fat, sugar and all the other nutrients for one Danish pastry at your fingertips.

World Cancer Research Fund

www.wcrf-uk.org

The WCRF is one of the leading charities in the field of diet, nutrition and cancer prevention. Their website contains heaps of cancer-related information including cancer-prevention guidelines, overviews of current research and free publications. As you'd expect given the links between obesity and cancer, the site offers great advice on weight loss including a review of the pros and cons of different weight loss diets. The site has a good section that discusses the links between red meat and cancer and a quiz to assess how healthy your diet is. We also like the 'What's in season?' section, which can help guide you to the best value fruits and vegetables at any time of year.

You can also find reliable, easy-to-understand patient information from Cancer Research UK on the website www.cancerhelp.org.uk. This website has a particularly useful section on how to cope with the dietary problems that may arise during cancer treatment, such as weight loss, swallowing problems or taste changes.

Chapter 26

Ten Superfoods

*A*ll foods contain valuable nutrients – yes, even a deep-fried Mars Bar has some goodness. However, some superfoods have so many health-giving properties that eating them on a regular basis could be a real bonus. Here are our top ten superfoods based on the latest nutrition research.

Apples

An apple a day keeps the doctor away, or so you've always been told. Over 7,500 varieties of apple are grown throughout the world and any one counts as one of your five-a-day portions. Like bananas, apples are packed full of antioxidants, especially vitamin C for healthy skin and gums – one apple provides a quarter of your daily requirement of vitamin C. In addition, apples contain a form of soluble fibre called pectin that can help to lower blood cholesterol levels and keep the digestive system healthy.

An apple is also a handy package of carbohydrate of the low glycaemic index (GI) type. Your body digests low GI foods slowly; after they're finally broken down in the intestine they're gradually absorbed into the bloodstream as glucose, causing a gradual rise in blood sugar levels. This means that your body needs to release only a small amount of the hormone insulin to keep blood sugar within the normal range. A regular inclusion of low GI foods such as apples may help to improve diabetics' long-term control of blood sugar levels. Low GI foods may also help with weight control and may protect against heart disease.

Bananas

Did you know that bananas don't actually grow on trees but on giant herbs? It's also a myth that bananas are fattening. Bananas are slightly higher in energy than other fruits but the calories come mainly from carbohydrate; excellent for refuelling before, during or after exercise.

Bananas also make a very useful contribution to the recommended daily five portions of fruit and vegetables. All types of fruit and vegetables contain plant chemicals or phytochemicals known as *antioxidants*. These antioxidants protect cells in the body against damage from free radicals that can cause heart disease and cancer. Increasing your intake to five portions of fruit and veg a day reduces the risk of death from these chronic diseases by about 20 per cent. Bananas are also jam-packed with potassium that helps lower blood pressure, and vitamin B6 for healthy skin and hair.

Baked Beans

The humble baked bean is a nutritional powerhouse of protein, fibre, iron and calcium. It contains carbohydrate that, like that in apples and seeded breads, is of the low GI variety.

The tomato sauce covering baked beans is also a good source of lycopene, another powerful antioxidant shown to help prevent heart disease and prostate cancer. Lycopene is found predominantly in tomatoes and appears to be more available to the body when tomatoes are processed or mixed with other foods. So the sauce with your beans is a lot more than a tasty accompaniment to make your toast soggy!

The insoluble fibre in baked beans isn't digested but moves into the large intestine, or colon, where bacteria act on it and produce short-chain fatty acids. These fatty acids are thought to nourish the colon lining and protect it from carcinogenic (cancer-causing) invaders. Insoluble fibre adds to the physical bulk of the stools, helping them to move along more rapidly. This helps prevent the cancer-causing substances from attaching themselves to the colon wall.

Brazil Nuts

All nuts are generally packed full of essential vitamins, minerals and fibre. Recent studies suggest that eating a small handful of nuts four to five times a week can help reduce heart disease, satisfy food cravings and even control weight. Unsalted nuts are best, because most people already get far too much salt from the rest of their food. Brazil nuts are one of the few good sources of selenium that may help protect against cancer, depression and Alzheimer's disease. Just two brazil nuts a day meets your selenium requirement.

Broccoli

The old advice to eat up your greens to stay healthy is true. Although the intake of many vegetables has gone down in the UK, broccoli has increased in popularity in recent years. Just two florets – preferably raw or lightly cooked – count as a portion.

Not only does broccoli contain antioxidants including vitamin C, but it's a particularly good source of folate (naturally occurring folic acid). This B vitamin has been found to reduce levels of an amino acid known as homocysteine in the blood. High homocysteine is linked to an increased risk of heart disease and is now believed to be as important a risk factor as high cholesterol levels. Nutritionists believe that increasing your intake of folic acid is of major benefit in preventing heart disease, and studies are currently underway to investigate further. Recent research has linked high homocysteine to an increased risk of Alzheimer's disease and osteoporosis as well.

Broccoli also contains a particular antioxidant called lutein that can delay the progression of age-related macular degeneration (AMD). AMD affects 10 per cent of people over 60 (and 20 per cent over 80) and is a major cause of impaired vision and blindness. So it's not just carrots that are good for eyesight. And as if all this good stuff isn't enough, broccoli contains another phytochemical called sulphoraphane that has specific anti-cancer properties.

Coffee

Coffee has a range of useful properties. The caffeine content is helpful for stimulating alertness, mood and motivation, especially when you're tired or run-down. When taking a break from driving, the Department for Transport's THINK Road Safety advice encourages you to have two cups of coffee or other high caffeine drinks alongside a short nap (no more than 15 minutes), to allow time for the caffeine to kick in, before continuing your journey.

And did you know that coffee counts towards the six to eight cups of fluid you're recommended to drink daily? At the typical levels that people drink coffee in the UK, you take in insufficient caffeine to suffer from any diuretic effect.

Despite negative reports in the media relating to coffee, a Finnish study of over 20,000 individuals found no association between coffee consumption and coronary heart disease (CHD). And an American study of 27,000 older women over 15 years found around a 30 per cent reduced risk of cardiovascular disease in women who had a moderate intake of coffee. Further analysis found that up to 60 per cent of antioxidants in their diet may have come from coffee. New areas of research looking at coffee's potential preventative role are focusing on conditions such as type 2 diabetes and Parkinson's disease.

Oats

Oats are wholegrains and as such are a good source of B vitamins, particularly thiamin, riboflavin, B6 and folate, as well as vitamin E. They're low in sodium and a good plant source of potassium. Oats are also a source of soluble fibre. The European Food Standards Agency claims that 'Regular consumption of beta-glucans contributes to maintenance of normal blood cholesterol concentrations'. And oats are low GI foods. They can therefore help to control blood glucose levels by reducing post-meal rises in blood glucose levels and providing a steady release of energy.

Try an oatcake or two. Depending on the recipe, oatcakes traditionally contain 78.4 per cent wholegrain, meaning you need to consume approximately two oatcakes to consume one of the recommended three daily 16-gram servings of wholegrain.

Olive Oil

Olive oil is prized in many parts of the world for its taste and versatility and is the cornerstone of many great cuisines. In ancient medicine people used the oil to cure pretty much everything from mental illness to ulcers. Several large studies suggest that the monounsaturated fat in olive oil is good for the heart. Olive oil lowers bad cholesterol levels and increases the good levels (see Chapter 5 for more on the goodies and baddies of the cholesterol world). Olive oil is also rich in antioxidants– in fact, it's probably one of the key protective aspects of the so-called Mediterranean diet.

You can use olive oil for cooking or drizzled on salad. Olive oil-based spreads are an excellent alternative to butter.Watch out for the calories – a little goes a long way: A tablespoon of oil contains 120 kilocalories: the same as a large slice of bread and butter. If you're not keen on the taste of olive oil, then rapeseed oil (often sold as blended vegetable oil) is nearly as good but just as high in calories.

Sardines

All fish is a source of good-quality protein, vitamins and minerals, but oily fish such as sardines also contains omega 3 fats that reduce blood clotting (thrombosis) and inflammation. Studies show that eating oily fish dramatically reduces the risk of having a heart attack, even in older adults, and it's never too late to start. Omega 3 fats have also been shown to have a myriad of other benefits. They help prevent depression, and protect against the onset of dementia. Yep, it's true: fish really is an all-round brain food.

Eat at least one serving of oily fish per week, increasing to two to three servings per week if you're at high risk of heart disease. Alternatives to sardines include pilchards, trout, herrings, mackerel, salmon and fresh tuna. Tinned (not tuna, because the omega 3s are destroyed during the processing of canned tuna), fresh and smoked fish are all good.

Yoghurt

To make yoghurt, you add a bacterial culture to warm milk. The bacteria feed on the milk sugar called lactose, and release lactic acid, which thickens the milk into the familiar creamy texture of yoghurt. Yoghurt is an easily absorbed source of calcium, and one small carton provides 40 per cent of an adult's daily calcium needs. It's also a useful milk substitute for people who can't digest large amounts of the milk sugar, lactose.

When you turn milk into yoghurt, the bacteria culture you add to the milk produces lactase that breaks down lactose, the milk sugar.

Yoghurt has long been credited with a range of therapeutic benefits, many of which involve the health of the large intestine and the relief of gastrointestinal upsets. At the beginning of the 1900s, the Russian researcher Metchnikoff won the Nobel Prize for his investigations into why people in some parts of eastern Europe live particularly active lives and stay in good health well into their 90s and beyond. He concluded that you could attribute this phenomenon to the yoghurt in their diet.

The bacteria *Lactobacillus GG*, added to some yoghurt, aren't digested, and reach the large intestine intact where they 'top up' the other friendly bacteria living there. The friendly bacteria fight harmful bacteria, including *Clostridium difficile* that can cause diarrhoea after a course of antibiotics. Scientists have since discovered other strains of friendly bacteria like *Lactobacillus immunitas*; many now appear in yoghurts and other functional foods designed specifically to improve gut health. So as well as choosing between strawberry or raspberry yoghurt, you can also decide which type of friendly bacteria you want to snack on!

Chapter 27

Ten Fad Diets: The Truth Behind the Headlines

. . . .

In This Chapter

▶ Checking out how different diets work – or not

▶ Exploding the dieting myths

. . . .

*Y*ou only have to flick through a glossy magazine to find one 'revolution-ary' diet plan or another. Claims of rapid, dramatic weight loss and the resulting 'essential' pencil-thin figure lead you to think that this latest diet is the miracle you're looking for! But just what's so special about the *en vogue* regimes used by the stars, and, more importantly, how effective are they?

The Atkins Diet

Named after the infamous American Dr Atkins, the Atkins company became a global multi-million-pound organisation but declared bankruptcy in the US in 2005.

Selling over 10 million copies worldwide and outdoing both *Harry Potter* and the Bible, the Atkins Diet is based on restricting carbohydrate intake to as little as 15 grams per day and indulging in high-protein and high-fat foods. Three important studies in 2003 showed that the first stage of the diet is consistent with most weight-loss plans, but in the long term the diet and weight loss are both difficult to maintain. The jury is still out on the long-term safety of the regime and nutritionists are concerned about the increased risk of nutrient deficiencies, constipation and ketosis (the breakdown of body tissues to pro-duce energy) as well as kidney and heart health in Atkins devotees.

Atkins seems to be falling out of fashion.

The 'Bit of Your Body You Hate' Diet

If you're unhappy with your hips, thighs, bum or boobs, a diet is out there promising to make them smaller, firmer and more toned. You can even follow a diet claiming to give you the same effect as a face-lift in a weekend! Most of these diets give you a rigid set of rules to follow, with a whole list of foods you must eat at certain times of the day in certain combinations to release the 'special' powers of the nutrients. These diets often advocate foods that have a high protein and vitamin content, providing a tentative link to improving your skin's elasticity or rebuilding new skin cells. Your body uses protein and vitamins to make new skin cells, but that's where the magic ends – no food can target a specific area of your body to work on. If you lose weight you may find your least favourite 'wobbly' bit becomes smaller, but the only way it will ever get firmer and more toned is through plain hard exercise.

Although most of these targeted diets aren't likely to do any harm, they're more of a marketing ploy to sell magazines and newspapers than a breakthrough in nutritional science. The basic principles of the diets are the same: reduce calorie intake and increase calorie expenditure through exercise to lose weight. But, of course, that doesn't make a very sexy headline!

A lot of hype and not much science here.

The Blood Group Diet

The blood group diet suggests that eating certain foods according to your blood type rids your body of toxins and fat. The diet varies for each blood group, so, for example, a person with an O blood type can eat organic seafood, red meat, soya milk and most fruit and vegetables, but cuts out wheat, potatoes, dairy products and some vegetables. However, little if any scientific research validates this.

Any weight loss on this diet is likely to be attributable to a simple reduction in total calories rather than any supposed physiological response to food related to blood type. Apart from leaving you hungry most of the time, the blood group diet is pretty expensive and can lack fibre, essential fatty acids and calcium unless you also eat fortified soya products.

The Cabbage Soup Diet

Most people spent their childhood loathing it, but now it's touted as one of the hottest ingredients in the world of wonder diets! The humble cabbage, made into soup and eaten every day in addition to fruit, vegetables and occasionally meat, ensures rapid weight loss due a very low calorie intake, rather than any magical properties of the cabbage. This diet is dangerous to maintain over a long period. The removal of a limb is preferable to this regime for some rapid weight loss!

Nutritionally incomplete and very much a quick fix: Detox Diets

Detox diets usually consist of fruit, vegetables, seeds and herb- or fruit-infused water with limited wheat-free grains and oils. Detox plans are often sold on the pretence of their 'cleansing health benefits', but in truth the resultant and often dramatic weight loss is down to simple calorie restriction and is usually short-lived. Health professionals have genuine concerns about people frequently coming on and off detox plans, or following them for any length of time. Detox diets are nutritionally incomplete and rapid weight loss can exacerbate the yo-yo diet effect (rapid weight loss, with equally rapid weight gain).

Detoxing for a day or two is unlikely to do you any harm, but it should never be a permanent solution to weight control.

The Hay Diet, a.k.a. the Hollywood Diet

The Hay Diet is named after its founders who developed the regimen to treat Mr Hay's cancer.

Food combining – the main tenet of the Hay diet – means eating carbohydrates and protein separately and only eating fruit before a meal or a couple of hours after. No scientific evidence suggests that food combining works. Some people say they feel better and do lose weight following the diet, but any resulting weight loss is likely to be due to the simple fact that people on the Hay diet tend to eat less food. In practical terms, following a food-combining diet is difficult to mix with eating out, socialising and feeding a family.

All in all, this diet is inconvenient and just plain hard work!

Low Glycaemic Diets

The glycaemic index (GI) and glycaemic load (GL) are tools to measure the rate at which the body absorbs carbohydrates as glucose into the bloodstream. The theory is that controlling fluctuating blood sugars by including more low or 'slow' GI foods like pulses, new potatoes and oats at mealtimes reduces food cravings and the high-calorie grazing that can lead to weight gain. Many health professionals support the use of low GI foods as part of a healthy diet and active lifestyle and some studies show that as well as helping weight control in the short-term, the risks of heart disease, diabetes and *metabolic syndrome* (combined obesity, raised cholesterol, blood pressure and impaired glucose tolerance) may also be reduced.

To get the full benefits of a low glycaemic eating plan, use it as part of the healthy diet and lifestyle we talk about in Chapter 14.

Low glycaemic diets are likely to be around for a long time so watch this space.

The Maple Syrup Diet

The syrup – Madal Bal Natural Tree Syrup – is made from the sap of maple and palm trees. Also known as The Lemon Detox, the diet was introduced more than 30 years ago by naturopath Stanley Burroughs. Many of the claimed benefits include weight loss, cleansing the body of toxins, greater resistance to illness, improved concentration, increased energy, clearer skin and eyes, shinier hair and stronger nails.

The detox drink itself is made by mixing 2 tablespoons (which the diet says is 20 millilitres) of the Natural Tree Syrup with 2 tablespoons of freshly squeezed lemon juice, a pinch of cayenne pepper or ginger and half a pint of hot or cold water. It's recommended you drink as much as you like.

This diet is potentially dangerous. Effectively, it involves drinking nothing but sugared water for days on end. Of course you'll lose weight, but that's purely the result of limiting calories excessively.

A tablespoon (which the diet says is 10 millilitres) of Natural Tree Syrup contains just 26 calories. That means if you have six drinks a day (each containing the recommended 2 tablespoons of syrup), your daily intake will be just 312 calories. Even nine drinks only provide 468 calories.

Meal Replacement Diets

Meal replacements are portion-controlled products fortified with vitamins and minerals. You replace one or two meals with these liquid shakes or bars, allowing one ordinary low-calorie meal a day. This approach provides an energy intake in the region of 1,200–1,600 calories per day.

Meal replacements are as effective as traditional dietary treatments in the short term, with long-term weight maintenance. None of the research published to date suggests any adverse effects of using meal replacement diets.

At present it isn't possible to predict who does best with this kind of approach. However, it seems a good option if you've tried to lose weight using more traditional dietary methods, or have difficulty finding time to prepare meals or understanding and controlling portion sizes.

The Zone

The Zone regime combines a low-carbohydrate/high-protein diet (like the Atkins diet) along with a rigorous, intensive exercise routine. You can eat anything low in carbohydrate and high in protein. However, over time, the combination of exercise and low carbohydrate intake leads to the breakdown of muscle stores, reduces performance and increases general fatigue. Competitive sports men and women swear by a diet high in carbohydrates to provide the fuel they need for their sport, so the Zone contradicts medical, nutritional and sports science. It's difficult to imagine how followers of this diet can maintain relentless exercise without the right kind of fuel and nutrients.

One to avoid!

Index

• L •

• R •

• S •

FOR DUMMIES®

Making Everything Easier! ™

UK editions

BUSINESS

Marketing Kit FOR DUMMIES

978-0-470-74490-1

Business Plans Kit FOR DUMMIES

978-0-470-74381-2

Consulting FOR DUMMIES

978-0-470-71382-2

REFERENCE

British Politics FOR DUMMIES

978-0-470-68637-9

Football FOR DUMMIES

978-0-470-68837-3

Researching Your Family History Online FOR DUMMIES

978-0-470-74535-9

HOBBIES

Growing Your Own Fruit & Veg FOR DUMMIES

978-0-470-69960-7

Allotment Gardening FOR DUMMIES

978-0-470-68641-6

Electronics FOR DUMMIES

978-0-470-68178-7

Anger Management For Dummies
978-0-470-68216-6

Boosting Self-Esteem For Dummies
978-0-470-74193-1

British Sign Language
For Dummies
978-0-470-69477-0

Business NLP For Dummies
978-0-470-69757-3

Cricket For Dummies
978-0-470-03454-5

CVs For Dummies, 2nd Edition
978-0-470-74491-8

Divorce For Dummies, 2nd Edition
978-0-470-74128-3

Emotional Freedom Technique
For Dummies
978-0-470-75876-2

Emotional Healing For Dummies
978-0-470-74764-3

English Grammar For Dummies
978-0-470-05752-0

Flirting For Dummies
978-0-470-74259-4

IBS For Dummies
978-0-470-51737-6

Improving Your Relationship For
Dummies
978-0-470-68472-6

Lean Six Sigma For Dummies
978-0-470-75626-3

Life Coaching For Dummies,
2nd Edition
978-0-470-66554-1

FOR DUMMIES®

A world of resources to help you grow

UK editions

SELF-HELP

978-0-470-66541-1

978-0-470-66543-5

978-0-470-66086-7

STUDENTS

978-0-470-68820-5

978-0-470-74711-7

978-0-470-74290-7

HISTORY

978-0-470-99468-9

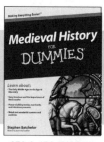

978-0-470-74783-4

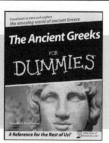

978-0-470-98787-2

Origami Kit For Dummies
978-0-470-75857-1

Overcoming Depression For Dummies
978-0-470-69430-5

Positive Psychology For Dummies
978-0-470-72136-0

PRINCE2 For Dummies, 2009 Edition
978-0-470-71025-8

Psychometric Tests For Dummies
978-0-470-75366-8

Raising Happy Children
For Dummies
978-0-470-05978-4

Reading the Financial Pages
For Dummies
978-0-470-71432-4

Sage 50 Accounts For Dummies
978-0-470-71558-1

Self-Hypnosis For Dummies
978-0-470-66073-7

Starting a Business For Dummies,
2nd Edition
978-0-470-51806-9

Study Skills For Dummies
978-0-470-74047-7

Teaching English as a Foreign
Language For Dummies
978-0-470-74576-2

Teaching Skills For Dummies
978-0-470-74084-2

Time Management For Dummies
978-0-470-77765-7

Work-Life Balance For Dummies
978-0-470-71380-8

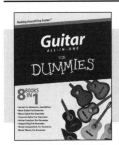

FOR DUMMIES®

Helping you expand your horizons and achieve your potential

COMPUTER BASICS

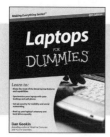

978-0-470-57829-2

978-0-470-46542-4

978-0-470-49743-2

DIGITAL PHOTOGRAPHY

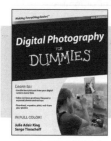

978-0-470-25074-7

978-0-470-46606-3

978-0-470-59591-6

MICROSOFT OFFICE 2010

978-0-470-48998-7

978-0-470-58302-9

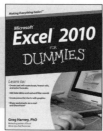

978-0-470-48953-6

Access 2007 For Dummies
978-0-470-04612-8

Adobe Creative Suite 5 Design
Premium All-in-One For Dummies
978-0-470-60746-6

AutoCAD 2011 For Dummies
978-0-470-59539-8

C++ For Dummies, 6th Edition
978-0-470-31726-6

Computers For Seniors For Dummies,
2nd Edition
978-0-470-53483-0

Dreamweaver CS5 For Dummies
978-0-470-61076-3

Excel 2007 All-In-One Desk Reference
For Dummies
978-0-470-03738-6

Green IT For Dummies
978-0-470-38688-0

Macs For Dummies, 10th Edition
978-0-470-27817-8

Mac OS X Snow Leopard For Dummies
978-0-470-43543-4

Networking All-in-One Desk Reference
For Dummies, 3rd Edition
978-0-470-17915-4

Photoshop CS5 For Dummies
978-0-470-61078-7

Photoshop Elements 8 For Dummies
978-0-470-52967-6

Search Engine Optimization
For Dummies, 3rd Edition
978-0-470-26270-2

The Internet For Dummies,
12th Edition
978-0-470-56095-2

Visual Studio 2008 All-In-One Desk
Reference For Dummies
978-0-470-19108-8

Web Analytics For Dummies
978-0-470-09824-0